Landing Your First Job
A Guide for Physics Students

John S. Rigden
American Institute of Physics

American Institute of Physics
Career Services Division

College Park, Maryland

www.aip.org/careersvc
csv@aip.org

August 2002

Published under the aegis of the American Institute of Physics
Career Services Division.
ISBN 0-7354-0080-6

ii

CONTENTS

Preface

Where do physics students go for help and advice when they anticipate starting a job hunt? Most likely, you would go to your physics professor. However, most academic physicists begin their employment on a campus and they end it on a campus; therefore, they may have little or no experience in industry or government which is where the great majority of physicists work. So your physics professors may be unable to provide the detailed advice and the focused guidance that students need as they plan their job search. You could go to your campus career center. And you should. The career center staff can be very helpful. In addition, read this short book. This book, *Landing Your First Job*, was written to guide students who are looking ahead to seek employment. It is a practical book that identifies the important steps to the first job.

**Freshmen, sophomores,
and first-year graduate students:
start now
to plan your job search.**

The information contained in *Landing Your First Job* has a number of origins. One valuable source of information was an outgrowth of the many career workshops organized and conducted by the Career Services Division of the American Institute of Physics. These workshops featured headhunters, directors of human resources, and others who work in the employment field. A second source of information was physicists employed in a variety of industrial and governmental positions. From their own experiences and those of others, they provided suggestions on getting the first job. The directors of human resources from several companies revealed their working strategies and identified what got their attention.

This book was produced by the American Institute of Physics in College Park, Maryland. If you have comments about it, please direct them to Career Services, American Institute of Physics, One Physics Ellipse, College Park, MD 20740 or by e-mail to csv@aip.org. Copies of this book may be ordered from the same postal address, e-mail address or by going to the web site www.aip.org/careersvc.

INTRODUCTION

This book is for undergraduate and graduate students of physics. By studying physics, you are building a strong foundation on which to base your future career. A bachelor's, master's, or doctor's degree in physics endows you with valuable knowledge and practical skills needed and respected in the workplace. So, for choosing physics, congratulations.

When you graduate with your degree in physics, you hope to segue smoothly into an interesting, challenging, and rewarding job. Any new degree-holder needs guidance as they prepare to enter the job market, but this is especially true for someone seeking their first "real" job and it is particularly true for someone with a degree in physics. Why is this? Simply put, getting the first job is harder than getting the second job. And the first-job challenge is especially keen for physicists because very few positions are explicitly earmarked for physicists. This means a physicist must adopt job-seeking methods that connect you, as a physicist, directly to the job you desire. This book identifies how this can be done. This book is about getting that first job.

One qualification. Your physics professors can tell you all about getting a job in academe; therefore, this book focuses on getting jobs in the wide world outside of academe: in industry, government, and elsewhere. If you want to become a physics professor, talk to your physics professors.

Physicists with baccalaureates, physicists with masters, and physicists with doctorates work in a rich variety of jobs, mostly in industry, and pursue a diverse range of technical and nontechnical careers. Everyone knows where a degree in dentistry leads to, but one cannot say this about a degree in physics. Since the principles of physics underlie all the technologies that drive the workplace, physicists do many jobs and, as experience verifies, they do many jobs well. Physicists work as technicians, engineers, managers, computer specialists, technical coordinators, designers, trainers, financial analysts, bosses, researchers, entrepreneurs, presidents, owners, consultants, and the list goes on. Given the diversity of careers pursued by physicists, you might want to expand your expectations about what you can do with a physics education.

> **Employers pay you
> to meet their needs
> and contribute to the bottom line.**

To get your first job, you will be functioning in a competitive environment. It is important to take steps, starting now, that will make you an attractive job candidate. It is important, starting now, to build a network of individuals who can give you good advice and help open doors for you. It is necessary for you to know how to present yourself and to practice doing it. Therefore, you will need to assess anew the knowledge, skills, talents, and abilities deriving from your education and your general experiences. This assessment identifies what it is you bring to the job market. Further, as you better understand your strengths, you will build your confidence, better understand what you want, and focus and refine your job search.

This book, *Landing Your First Job*, is designed to identify and help you implement the steps to your first job. In the following pages you will find practical information for your job search including: networking, self assessment, targeting potential employers, learning about potential employers, writing cover letters, creating a resume, preparing for an interview, and negotiating compensation. Appearing in this book are statistical data that can guide your thinking and decisions. Throughout the following pages we attempt to be as physics-specific as possible; however, at times the material is more generic. The material is generic in another sense. This book is directed to both undergraduate and graduate students; however, since almost half of all baccalaureates in physics immediately enter the job market, more references are made to the undergraduate student. When differences in job-search approaches exist between undergraduate and graduate students, for example, resumes vs. CV's, distinctions are made.

To land your first job, be prepared to work.

Let us hear from you as you go through the process; tell us your difficulties and successes. Tell us about getting your first job and how your career evolves. Perhaps your own experiences will be featured in future editions of this book. Give us your comments about the book. Tell us what is useful and what is not, what is missing and what you would like more of. Above all, tell us how this information helped you in seeking your job. We would like to hear your story. Once again, the e-mail address is csv@aip.org.

This book will help you develop a strategy for finding your first job.

After that, it is up to you.

Steps to a Job

Before the search

STEP 1: Make decisions as a freshman, sophomore, or first-year graduate student that will strengthen your resume and attract attention to you as a job candidate.

STEP 2: Establish a network of contacts among the alumni of your department and the soon-to-be alumni, the advanced students in your department.

Preparing for the search

STEP 3: Know yourself. Spend the time necessary to do a careful self assessment.

STEP 4: Make appointments for informational interviews at two or three companies. These need not be places where you want to seek employment. The purpose is to broaden your knowledge of the workplace.

STEP 5: Target two or three potential employers that interest you and study them in depth to learn and understand what they do and how they generate their revenue.

STEP 6: Expand your network to include individuals who are employed by your targeted employers or who know people who work for your potential employers. Talk to members of your network about these employers. Quiz them about each employer.

The search

STEP 7: Through conversations with members of your network, by examination of your targeted employer's website, or through electronic or printed help wanted ads, determine the position (or positions) you are going to apply for.

STEP 8: Design each cover letter and resume so that it is targeted toward the specific position you desire. This means using keywords identified with the position and highlighting your knowledge and skills required to fill the position.

STEP 9: Prepare for interviews. Again, be prepared to connect your abilities with the demands of each specific job.

STEP 10: Be ready to negotiate your compensation.

To Freshmen, Sophomores and 1st- Year Grad Students: Laying the Foundation

> ## Step 1
> **Make decisions as a freshman, sophomore or first-year graduate student that will strengthen your resume or CV and attract attention to you as a job candidate.**

When you send your resume and cover letter to your would-be employer, your student days will be mostly behind you. The strength of your resume will depend on choices you have made one year, two years, or three years earlier. The product of those choices will be a foundation from which you will launch your job search. The better your choices, the better your foundation. The better your foundation, the better your chance of getting the job you desire.

Many physics students share a common image of "the physicist" which is at odds with the facts. It is commonly assumed that a physicist is a Ph.D., a physicist does basic research, and, most often, does so at a college or university. This image is inaccurate. Physicists with baccalaureates, masters, and doctorates pursue widely diverse careers and they do many things other than basic research. The companion to this book, *Physicists at Work*, introduces students to the variety of jobs and work environments of typical physicists.

So, where do physicists work? If we define physicist as someone with a bachelor's, master's, or doctor's degree in physics, then in broad-brush terms, here is an approximate answer:

> ~ 54% in industry, national laboratories, and government
> ~ 21% pursue graduate studies in disciplines other than physics (mathematics, chemistry, engineering, computer science) and in professions such as medicine, law, and business administration
> ~ 16% uncertain
> ~ 5% in departments of physics

The lesson to be taken from these data is that a physics student should anticipate the likelihood of working in industry or government, in other disciplines, or in one of the professions. If you make good choices, you will greatly enhance your attractiveness to all potential employers and thereby maximize your options.

> **Make choices today that maximize
> your options tomorrow.**

Lay a strong foundation for what? The long answer to this question is this: For your professional career. The short answer is: For getting your first job. A degree in physics, in and of itself, is a strong foundation for the long term. With a physics degree as the starting point, physicists successfully pursue a wide range of careers. That speaks to the foundational strength of physics. For the short term, however, there are important ways to enhance the foundation physics provides.

Advanced Preparation for the Job Search

Early as a physics student, undergraduate or graduate, long before you begin thinking about particular jobs, do three things:

> 1. Take one burden off your back,
>
> 2. Make the acquaintance of alumni from your department,
>
and
> 3. Make choices that will set you apart.

Remove One Burden and Relax...a Bit

What three things dominate the typical student's thoughts as he or she pursues an academic degree? Regardless of the discipline, three things? Here's a guess.

> First, getting out of the nest, becoming your own person, and lessening or ending Mom's and/or Dad's control. (This is particularly true on commuter campuses where attending college and living under Mom and Dad's roof can create real tensions.)
>
> Second, sex. Nothing more need be said.
>
> Third, deciding what you're going to do with your life - particularly with regard to employment and career.

Each of these is a nitty-gritty issue that competes for students' attention as they try to read Plato's Republic or calculate the electric field of a cylindrical charge distribution.

As far as Moms, Dads, boyfriends, or girlfriends are concerned, you are on your own. It doesn't mean these issues are not important because they are. The third issue, however, employment and career, is where I offer advice. If I were to reduce our advice to one word it would be RELAX. Let me explain.

Often students believe they must decide today what they will be doing 10 years from today. That belief introduces a major burden and it is largely unnecessary. Of course, you should think long range, but do so in terms of broad principles. For example, I want to work in a technical environment, I want to work closely with other people, I want to serve the needs of society, I want to apply knowledge to practical ends, I want to make big money, etc. Such principles guide your thinking, but they cannot reveal where you will be working a decade hence.

Beyond the broad principles, the reality is this: any decision made today about 10 years hence will likely be negated by a whole series of events along the way. The first job, whatever it may be, will introduce unanticipated variables that will significantly influence how you think about your next career step. During your first job, your circle of professional contacts will point out new opportunities. You cannot predict this. New job offers will come. You cannot predict this. Technological changes can alter the workplace in ways that define new directions. You cannot predict this. Personal and family considerations may prompt unplanned changes. You cannot predict this. Conclusion: don't waste a lot of time and energy trying to map out your life's career. Put your energy into thinking directed towards your first job.

A job is not a career.

A career spans a lifetime.

During your first job, you become better equipped to think long range.

De-emphasize 10 years from now;

Emphasize now and next year.

Networking I

Learn about Alumni

Step 2
Establish a network among the alumni and the advanced students of your department.

The alumni from your department, once physics students like you, are natural allies. They know what you are going through as a physics student, they have the experience under their belt of finding their first job, and they are currently working in a variety of positions for a range of employers. Get to know the physics alumni and begin to build a network of individuals who can help you. Of course, the advanced students will soon be alumni so with them, start early.

Organize a group of physics students - undergraduate and/or graduate - who know that one day they will be starting their own job search. Get a list of physics alumni from...someplace! The physics department *should* have such a list, but it probably doesn't. (If it doesn't, urge them to create such a list.) The alumni office will have some information (after all, they like to ask alumni for money), but its information may not be broken out by academic major. Visit your campus career center as it often has extensive alumni networks. However you do it, make the effort to get together a list of physics alumni with accurate addresses, e-mail addresses, and telephone numbers. The payoff can be great.

| **Today's junior and seniors are tomorrow's alumni. Get to know the upperclassmen.** | **Members of the physics alumni are your natural advocates.** |

As soon as you identify the whereabouts of alumni, invite them, one by one over time, to visit the department and talk to your student group. (You should tell faculty what you are doing. There may even be some money to help pay travel costs.) Experience shows that the alumni will most likely be delighted to receive such a request and will happily accept your invitation. If they are unable to visit you in the physics department, you might ask if a group of students could make an appointment to visit them at their place of employment. One way or another, find out about the actual work of various alumni and listen to the advice they offer.

As you learn about different alumni, organize your findings and save it for future reference. When you begin your job search, these alumni can be very helpful.

> **When I graduated, I wanted to get into high-tech sales. This was based on recommendations of friends and hi-fi colleagues. My first job was with Digital Equipment Corporation (DEC). I got it through the recommendation of an alumnae. DEC was looking for sales people with engineering degrees. My physics/math was close enough and I aced the interview partly due to my amateur knowledge of electronics.**
>
> Neel J. Price, Director Government Operations,
> Cetacean Corporation

Visit the Campus Career Center

Your campus career center has many resources that you should know about and, at the appropriate time, make use of them. Know what resources are there and know the services offered. Introduce yourself to a career counselor and talk with her/him about your plans as you currently see them. Ask for his/her advice. Make the career center staff part of your network. If you can get to know a career counselor so that you are more than just a name, the counselor can be valuable to you in many ways. Often career centers organize campus job fairs which bring a number of employers together to talk with and interview students. If a job fair is held on your campus or near by, take advantage of it.

> **I would consider college career placement centers and university-sponsored career fairs to be the best opportunity for college students.**
>
> Sharon Lappin, Project Engineer,
> Missouri Department of Transportation

Plan Now To Set Yourself Apart

Eventually, you will start looking for a job. During this process, you will find a position you really want. The employer is great, the location is ideal, and the position itself brings together responsibilities that are challenging and fulfilling. And the salary is attractive. You apply for the job...and hope.

Since the job you will seek is a good one, there will be other qualified candidates applying for the same position. They may be other physicists and almost certainly engineers. Physicists and engineers often compete for the same jobs. Just like you, all first-time job applicants

- will have a degree,

- will have taken relevant courses,

- will have earned respectable grades,

- will have developed good computer skills, and

- will likely have good recommendations or references.

So what sets **you** apart? What about **your** resume or CV attracts attention?

> **As a physics student in today's world, we assume you are developing computer skills. If you are not, do so.**

> **There are some interns who sit at their computer waiting for something to do. They don't ask questions or talk to anyone outside the department.**
>
> Judy Steele, who hires interns for Express Personal Services, and says she only considers hiring those who treat their internships like a real job and show initiative.

Following are things you can do - should do - that will absolutely strengthen your resume, attract attention to you as a job candidate, and may actually get you hired.

Do an Internship

A summer or semester internship (or co-op) takes your learning out of the classroom and exposes it to the realities of a work environment. An internship, for both undergraduate and graduate students, gives first-hand experience in the workplace. In addition, an internship will

- hone certain skills which you can highlight on your resume,

- expand your network with new acquaintances who can give you good advice and perhaps serve as employment references,

- enhance your understanding and appreciation of physics (the classroom is not a real-world environment),

- give you experience working with others (teamwork), and

- attract attention to you as a job candidate and prompt employers to give you a closer look.

An internship, in and of itself, often leads to a job offer.

> **We treat physics baccalaureates with an industrial internship preferentially.**
>
> Senior Physicist - Intel

Arrange For a Shadowing Experience

You could learn a great deal about a particular industrial work environment if you arranged to shadow an employee for one or two days. In other words, with a notebook in hand, follow someone around and observe them closely. What does she/he do? How is time allocated? What types of problems does he/she cope with? What is the nature of deadlines? Who does he/she work with? How do they work together? Take good notes. Discuss them with the individual you are shadowing. Keep an account of what you observe, put this experience on your resume, and take your notebook to your job interview.

Do Independent Research

Independent research under the direction of a faculty member is a substitute for an internship. This is especially true if (1) the research has an applied character, (2) the research is off-campus, and (3) you are able to describe in concrete ways how your research expanded your abilities, skills, and awareness of how physics can be applied.

Get an internship in industry.

Demonstrate Your Ability to Work with Others

In industry and government, professionals rarely work in solitude. To develop products, refine them, and get them out the door requires the skills of individuals from many different backgrounds. People work in teams to accomplish the desired objectives. The composition of the workforce making up these teams is increasingly diverse, sometimes with an international makeup. This means you will work with people from different religious, ethnic, cultural, and educational backgrounds. Create opportunities to work with others and thereby demonstrate your ability to be amiable and to work cooperatively.

Develop Your Communication Skills

Take definitive steps to develop writing and speaking skills. The importance of good communication skills cannot be overemphasized. A chronic complaint of leaders in all sectors of the workplace is the sorry state of the writing and speaking skills of their employees. Technical staff - bachelors, masters, and doctorates - must communicate with their managers and the managers must be able to read what you write and understand what you say. A graduate degree is, in and of itself, no guarantee of writing skills. Graduate students: learn to write jargon-free descriptions of your research work.

> **Working physicists put communication first or second in their lists of needed skills.**

As a physics student, you know that to learn the ideas of physics you have to work at it. Likewise, to acquire writing and speaking skills you have to work at it. More to the point, you have to **practice** writing and speaking. Take required English courses seriously. Take an additional writing course that will give you feedback. You need feedback about your writing! Understand why the instructor's red marks appear on your essay and learn from those red marks. If a course in science writing is available, take it. If there is a School of Journalism, enroll in its science writing course. Writing skill is very important for graduate students who will likely be writing proposals and reports as a part of their jobs. After courses, write, write, and write some more. Practice and practice some more. Write an essay on black holes, on lasers, or on the impact of physics on society and ask family members and friends to critique it for you. Do they **understand** what you are trying to say?

Here are some specific things you can do to develop your communication skills:

- as stated, take a writing course. Perhaps better, take a science writing course.

- organize a small student group and share your writing samples among yourselves. Critique each other.

- create a portfolio of your writing samples.

- form an interest group among a few friends and give short talks to each other. Critique each other's talk.

> **My physics education has been and still is the major driving force of my life, but, looking back, I wish I had taken some courses in accounting, finance, writing, and management.**
>
> Richard Buist, Owner and President
> TE Technology, Inc.

If your resume contains concrete evidence that you have worked to develop communication skills, this will demonstrate your recognition that communication skills are important and attract attention. If you show your portfolio of writing samples at your interview, you will make an enormous impression.

Finally, and without elaboration, here are other things you can do to enhance your resume and attract attention to your job application:

- become proficient in a second language (most businesses have customers overseas), and

- take a business course, for example, Introduction to Business or Introduction to Management.

To put in place those things that will enhance your value to a would-be employer, you must begin planning early. That is why this first chapter is addressed to freshmen, sophomores, and juniors. The pay-off for those efforts that set you apart from the rank-and-file job hunter will be huge.

The right things done now, will pay off later.

So, start now. Whether you are an undergraduate or graduate student, the best time to start preparing for a job is now. Throughout this book you will find information on what to do when looking for a job. There are several things, however, that you, as a student, can do right at this moment. Below is a list of "Action Points," some of which are drawn and adapted from the book *Careers In Science and Engineering* (National Academy Press, 1996). These action points are suggestions designed to stimulate ideas and *action*. Other action points will appear later. Select some of these and follow through.

Remember: If you have nothing to show an employer, the employer will have nothing to show you.

I started out with nothing.

I still have most of it.

Michael Davis, on the Tonight Show

Start now. Obtain and maintain the upper hand.

Looks like the upper hand

is on the other foot.

Lloyd Bridges in Hot Shots! Part Deux

ACTION POINTS I[*]

_____1. As an undergraduate or graduate student, work with your faculty advisor (or another faculty member) to plan a well-rounded education. Talk with faculty and students in professional programs. (Adopt an informal advisor in the humanities or the School of Business.)

_____2. Discuss potential academic programs with students and faculty, and via internet bulletin boards.

_____3. At both the undergraduate and graduate levels, take courses outside your major and primary field that you think will be useful in your career. (Another way to think about this: elect courses that will make you attractive to an employer.)

_____4. Seek advice from people outside physics as well as inside it.

_____5. Arrange an off-campus internship that can extend or broaden your skills and introduce you to another environment.

_____6. As an undergraduate, try to join a research team.

_____7. Look for classes and internships that will increase your breadth of experience.

_____8. Start writing. Create a portfolio and save writing samples.

_____9. Form an interest group; give talks to each other and encourage honest comment.

_____10. Volunteer for communication and leadership activities, e.g., in a disciplinary society, student organization, class or laboratory.

_____11. Build your confidence.

* Adapted from _Careers In Science and Engineering_, National Academy Press, 1996

From an early age, I have always excelled academically. From most academic in elementary school, to best history student in middle school, to attending a center-of-excellence school in high school, I have always had a strong propensity for learning. But in high school, my small town knowledge and motivations led me to join the military instead of going to college. To get away from home for the longest possible time and still remain a military reservist, I qualified for one of the longest technical schools in the Marine Corps. I became a Ground Radio Technician and worked on military radios and equipment. During military school, I ran track for my military base at several universities. It was at University of California, Northridge that I developed an interest in attending a university. I just could not believe that a place had such plush, green grass. I do not know why this caught my eye, but it did.

After military school, I returned home in North Carolina. Since a well-known university was just minutes away, I decided to apply in the only field that I knew anything about, vocal performance. (I had had formal training in high school as part of a chamber choir.) After a couple of semesters, I knew that the performing arts was not my cup of tea, so I changed my major to the only other area of academics that I had been exposed to, that is, electronics technology. From my formal training in the military, I was much more advanced than other students at my same grade level. Since I was working part time in my hometown, running cross-country for the university, and traveling 15 miles to and from school each day, I asked for a little financial assistance just to help me get my books before November. I was turned down flat for that assistance even though I was a top performer in that department. After my rejection, I went to my college physics class and, as usual, sat in the front. My the professor asked me why I was so dejected and I told him. This professor told me if I changed my major to physics, he would get my tuition waved and look into getting a little extra funding to help me with my educational expenses. Soon after, I was an engineering physics student.

As an engineering physics student, I had a flare for the electrical side of physics. I graduated top physics student of the department, a top electrical engineering student, and a well-known student teacher/tutor for general physics labs and lectures. My academic success and a couple of internships at well-known academic institutions and my background in military electronics made me very attractive as an entry- level electrical engineer for Raytheon. At Raytheon I started as an analog and digital design engineer and the first years were very intense and challenging. These first years set the tone and pace for what followed. After leaving Raytheon for an opportunity at Lucent Bell Labs, I began changing the direction of my career to technical project management. I found that my military leadership, student teaching, vocal performance for various events (big and small), and my rigorous training at Raytheon prepared me to lead technical projects. Because of my successes, companies would court me to join their team. My present employer was one such company and offered me a chance to help it grow. This is my dream and I will never stop searching for growth and opportunity.

Lee Phillips, Senior Product Engineer,
Broadband Development
ON Semiconductor

PHYSICS TRENDS

Who's Hiring Physics Bachelors?

For a listing of many of the companies that recently hired new physics bachelors in your state, see

www.aip.org/statistics/trends/emptrends.htm

These lists may be useful to job seekers in identifying the variety of companies that hire physics bachelors and to physics departments wishing to strengthen contacts with their local industry.

Statistical Research Center
www.aip.org/statistics

PHYSICS TRENDS

What Do Physics Bachelors Do?

Type of Job	Percent
Software	28
Engineering	17
Science & Lab Technician	10
Management, Owner & Finance	19
Education	12
Active Military	5
Service and Other Non-Technical	9

Type of employment of physics bachelors 5 to 7 years after earning their degrees, 1999.

Source: 1998 Bachelors Plus Five Study

AMERICAN INSTITUTE OF PHYSICS

Statistical Research Center

www.aip.org/statistics

THE FIRST-TIME JOB SEEKER

The first job may be the beginning of a career. Or it may not. A job and a career are two different things. Jobs and careers have different time scales. While there are exceptions, jobs have time scales of years; careers have time scales of decades. Jobs are defined from the beginning; careers unfold in slow and unpredictable ways and are not defined until the end. A job is a job; a career is a sequence of jobs.

The first job is not a career.

When you begin your search for that first job, recognize that your first job does not determine your career. This recognition will relieve some of the burden that attends job hunting. At the same time, your first job changes the rules on the playing field. You approach your first job with your college transcript and a resume identifying relevant experiences; you approach your second job from your first job. Once you have had a job, your student days - courses, grades, activities - become less significant. So, the first job is not a career, but it influences what follows.

One more suggestion. Time gaps in resumes attract the wrong kind of attention. What was she doing for three months between her first and second job? Why did he do "nothing" for six months after his graduation? These questions can nag at the minds of potential employers and can shift attention from you to another candidate. Sometimes students decide to reward themselves and take a break after graduation. By no means is such a decision fatal, but it can complicate the process.

Begin your job hunt before you graduate.

Learn all you can in your first job.

Success on the first job ups the ante for the second.

PREPARING TO SEEK YOUR FIRST JOB

Job seeking is not a casual activity. Particularly seeking the first job. It can be stimulating, exciting, and consciousness-raising. It can also be frustrating and stressful. Careful planning and preparation are necessary precursors for the job seeker. When should your planning begin? As early as possible - as a sophomore or 1st-year graduate student (see previous chapter).

If you set out on a job search with no principles to guide you, you can go in circles. To save hours, and possibly months, of aimless job hunting, you should spend some time right now to define yourself and your immediate goals. In other words, do a self assessment. Thinking and planning are everything. The following section, Self Assessment, is designed to stimulate your thinking and to help you decide what is best suited for you when you begin the job search.

SELF ASSESSMENT

> ## Step 3
> ## Know yourself.
> ## Do a careful self assessment.

What Do You Want Written on Your Tombstone?

I. I. Rabi was one of the famous physicists of the 20th century. When he was a couple of years shy of his 90th birthday, the *New York Times* sent a reporter to his New York apartment to interview him. This was nothing unusual as Rabi was often interviewed. But Rabi sensed this time was different and at one point he interrupted the reporter and said, "I know, you are interviewing me for my obituary." The reporter said, "Yes, Professor Rabi, that's true." Rabi responded quickly, "Can I read it?" Very gently, the reporter answered, "No."

Earlier we stressed the importance of writing. Here is something you could write. As we have said, deciding where you will be and what you will be doing 15 years from now is an iffy activity; however, thinking about what you want to accomplish by the end of your career is a useful exercise. So think a bit. As you consider yourself right now and as you look over the long years ahead, what do you want your legacy to be? What do you want to be remembered for? Or, to pose it in ghoulish fashion, how would you like your obituary to read? What accomplishments would you like noted? A good way to gain insight into yourself is to write your obituary as you would like it to appear. It's an odd assignment to think about your life's legacy while you are still a student, but there are personal questions, big questions, you can ask yourself.

> **It may seem silly
> to write your own obituary.
> In fact, however, you are doing it every day.**

If you don't fancy writing your own obituary, make a list of ten things you want to accomplish in your life. Such an exercise can make concrete the vague notions floating around in your head.

> **If you know what you want to do,
> you will better recognize the opportunities to do so.**

Self Assessment Game Plan

The objective of these pages is to know yourself. There are many ways to examine one's self. Probably no one way works well for everyone. So, four different approaches to self assessment follow.

- First, a series of questions designed to stimulate a particular line of reflection may help you assess yourself.

- Second, various skills are identified. From these lists, you can identify your own skills.

- Third, a test on work-related values with a weighting scheme can provide some indication of what you value in a job.

- Fourth, reflect on a list of intellectual and professional assets identified by physicists in industry.

Each of these assessment tools takes a different cut through the challenge of self assessment.

Questions for Self Assessment - I

? Do I have overarching goals?

> For example, do you want your work and/or your life to be recognized for its service to humanity and/or society? Do you want to acquire a national reputation and the opportunities/responsibilities that go with that? Do you want to earn big money?

? Do I want my life's story to be told through my own accomplishments or through my influence on others?

> A manager (or teacher) can be extremely effective in stimulating others to high levels of achievement while he or she is essentially invisible. Such managers take great satisfaction in the success of those under their supervision.

? Do I want my personal accomplishments to have tangible form?

> Some professionals are very successful, but have nothing to point to. Others feel the need to create a tangible legacy: books authored, patents obtained, instruments created, products developed, etc.

? Do I want to be a generalist or a specialist?

> Some enjoy being a jack-of-all-trades. They like variety and are generalists. Such people can make valuable contributions. Others find satisfaction in digging deeply into something and learning all there is to know about a particular subject. The latter becomes recognized as an expert.

There are other questions, besides those above, whose answers can reveal things deeply significant to you personally. As a student, soon to be seeking your first job, it may seem premature to think about your lifetime accomplishments. And it *is* premature. No immediate decision you are likely to make is going to write your life's script. At the same time, you make mundane decisions within an intellectual framework. The better you understand yourself, particularly regarding the big questions, the more accurately your intellectual framework will influence your decisions.

Questions for Self Assessment - II

? What do I enjoy doing?

? What are my personal strengths/weaknesses?

? What are my technical skills and experiences?

? What are my non-technical skills and experiences?

? Am I a good starter but weak at follow-through?

? Am I more a leader or a follower?

? Am I an idea person or a detail person?

? Am I a people person?

? Do I prefer a task where I work alone or with others?

? What is important to me - Money? Job Satisfaction? Prestige?

? What do I see myself doing in five years?

? What work sector - industry, government, non-profit - is of interest to me?

? Am I willing to relocate?

? What types of positions or responsibilities/duties are not acceptable?

? What are my salary needs?

Some of these questions are straightforward, others are not. An answer to one question can conflict with the answer to another. For example, an overarching goal to serve societal needs might conflict with salary considerations. An enjoyable job may not carry the desired prestige. Taken together, however, complete and partial answers to such questions will provide guidance during the job-hunting process. Use your answers to guide you to "the right job."

After getting a BS in physics and an MA in astronomy, I knew research wasn't for me. I looked for a job in science communication and was fortunate to get a position at Brookhaven National Laboratory. Now I use my computer and writing skills to keep the public up to date on scientific activities at this laboratory. My advice to anyone looking for an alternate career is to objectively evaluate your skills and strengths, then aim for jobs that use those strengths. Above all, enjoy what you do.

Christine Lafon, Physicist,
Brookhaven National Laboratory

Skills

This section is more specific - a focus on skills. There are, of course, generic skills that are the staple of all jobs. Then there are skills that are job-specific. Particular job environments often call for a particular range of skills. As you think about various job possibilities, identify what the job entails and what skills will be required to carry out the work successfully.

Generic Skills or Survival Skills

Communication: We talked about the communication issue earlier, but it is so important that we have come back to it. Communication skills are critical for obtaining a job and for doing it successfully. If you can convey your thoughts and describe your accomplishments, your attributes and your strengths, if you are able to describe why you are interested in a specific job at a particular company, an interviewer will respond more favorably to you as a job candidate. Effective communication requires the ability to organize thoughts, express those thoughts clearly, listen carefully, and show empathy for the lives and interests of others. Good verbal skills make it easier to find employment, to work with others, and to learn. One more thing: sometimes incorrect grammar will be overlooked; sometimes it won't. Good grammar counts.

> ## Good communication skills are as important as a specific technical competency.
>
> Jeff Schoenwald, Chief Physicist, Lynx Photonic Networks

Teaching: An important asset for finding a job in any sector is the ability to teach. When you interview, you are effectively teaching your interviewer about yourself. Whatever your job, you will find yourself teaching or coaching others.

Teamwork: Working with others requires "people skills": respect, understanding, compromise, etc. Teamwork comes naturally if you enjoy collaboration. In all jobs you will find yourself working with managers, administrators, committee members, and planners.

> **Teamwork**
> **Since receiving a MS in physics,**
> **I have worked in the digital image processing arena,**
> **along with other physicists, electrical engineers,**
> **image scientists, and mathematicians.**
>
> Teresa M. Froncek, Physicist, Eastman Kodak

Leadership: Some leaders are recognized formally as leaders; others are recognized informally. Leadership boundaries vary enormously - from an entire corporation to a co-worker. However, to some extent, everyone is a leader.

All of these generic skills can be improved through practice.

Specific Skills

Your physics education has focused primarily on building a knowledge base and honing some technical skills. Experience shows that this background leads to success in the workplace. However, there are other skills you have acquired (probably without knowing it) that are just as important as the technical skills. These skills include: the ability to reason, to spot interesting problems, to come up with problem solutions, to formulate hypotheses, and to test those hypotheses.

The American Institute of Physics has conducted frequent surveys asking working physicists to identify the skills they used most frequently in their jobs. Consistently, the skills ranked the highest are problem solving, interpersonal, and communication.

In the book, *Careers in Science and Engineering* published by the National Academy of Sciences, there is a list of personal qualities beneficial to the job search. These qualities are closely related to skills:

- intelligence, ability to learn quickly
- ability to make good decisions quickly
- analytical, inquiring, logical-mindedness
- ability to work well under pressure and willingness to work hard
- competitiveness, enjoyment of challenge
- ability to apply oneself to a variety of tasks simultaneously
- thorough, organized and efficient
- good time management skills
- resourceful, determined and persistent
- imaginative, creative
- cooperative and helpful
- objective and flexible
- good listening skills
- sensitive to different perspectives
- ability to make other people feel good about their work

The majority of these skills are included in the previous self assessment activities. Look back at your reactions and see if any of the personal qualities, important to employers, are also important to you.

Skills Most Desired by Employers

Learning to learn - the ability to absorb, process and apply new information quickly and effectively

Three R's - Reading, Writing, and Computation

Communication - The ability to communicate and listen effectively

Adaptability - Creative thinking and problem solving

Personal Management - Self-esteem, motivation/goal setting and career development/employability

Group Effectiveness - Interpersonal skills, negotiation and teamwork

Organizational effectiveness and leadership

To market yourself,
you must know yourself.

Work-related Values Test[*]

What sort of job do you really want? Where do you want to work? What environments do you prefer? In other words, what do you really value? Following are work-related values presented in the form of a test. Here is the grading scheme:

1 = **Not important at all**
2 = **Somewhat, but not very important**
3 = **Reasonably important**
4 = **Very important in my choice of career**

_____*Work alone:*   Do projects by myself, without any significant amount of contact with others.

_____*Independence:*   Be able to work/think/act largely in accordance with your own priorities.

_____*Knowledge:*   Engage myself in the pursuit of knowledge, truth, and understanding or work on the frontiers of knowledge, e.g., in basic research or cutting-edge technology.

_____*Expertise/Competence:*   Being a pro, an authority, exercising special competence or talents in a field, with or without recognition.

_____*Creativity:*   Create new ideas, programs, organizations, forms of artistic expression, or anything else not following a previously developed format (specify type of creativity).

* Margaret Newhouse, *Outside the Ivory Tower*

_____*People contact:* Have a lot of day-to-day contact with people - either clients or the public - or have close working relationships with a group; working collaboratively.

_____*Friendships:* Develop close personal friendships with people as a result of my personal work activities or have work that permits time for close personal friendships outside of work.

_____*Change or variety:* Have work responsibilities that frequently change in content and setting; avoidance of routine.

_____*Job stability/security:* Have predictable work over a long period and be assured of keeping my job and a reasonable salary.

_____*Career advancement:* Have the opportunity to work hard and make rapid career advancement.

_____*Wealth or Profit:* Have a strong likelihood of accumulating large amounts of money or other material gain.

_____*Recognition/Prestige/Status:* Be recognized for the quality of my work in some visible or public way; be accorded respect for my work by friends, family or community.

_____*Challenging problems:* Have challenging and significant problems to solve.

_____*Job pressure/Pace:* Work in situations with high pressure to perform well under time constraints; fast paced environment.

_____*Physical challenge:* Have a job that makes physical demands that I would find rewarding.

_____*Excitement/Adventure:* Experience a high degree of (or frequency of) excitement in the course of my work; have work duties that involve frequent risk taking.

_____ *Social service:*   Do something to contribute to the betterment of my community, country, society, and/or the world.

_____ *Services:*   Be involved in helping other people in a direct way, either individually or in small groups.

_____ *Moral fulfillment:*   Feel that my work contributes significantly to, and is in accordance with, a set of moral standards important to me.

_____ *Aesthetics:*   Be involved in studying or contributing to truth, beauty, culture.

_____ *Location:*   Find a place to live that is conducive to my lifestyle and affords me the opportunity to do the things I enjoy most or provides a community where I can get involved.

_____ *Power/Authority:*   Have the power to decide course of action, policies, etc. and to control the work activities or affect the destinies of other people.

_____ *Influence:*   Be in a position to change attitudes or opinions of other people.

_____ *Self Realization/Enjoyment:* Do work that allows realizing the full potential of my talents and gives high personal satisfaction and enjoyment.

These items help identify what you value. Group your responses by their rating to reflect on the results.

Communication skills

cannot be overemphasized.

Intellectual and Professional Assets for the Workplace

Each item below is identified as an asset for physicists in the workplace. Some of these workplace assets you may be able to identify as personal assets, but others you will meet in the workplace itself. The latter are included to give you a sense of what's ahead.

1. Reasoning from fundamental principles
   - ability to analyze and solve problems in a systematic way
   - not restricted to 'cookbook,' existing formulas, or 'rules of thumb'

2. Hierarchical precision
   - possess a feel for the scales and relative importance of different effects
   - ability to make first estimates, to conceive of simple and detailed models

3. Mathematical and numerical modeling
   - can describe physical phenomena quantitatively
   - can identify important variables and investigate the influence of many variables
   - can work with different design options and optimize designs

4. Multidisciplinary understanding
   - recognize the interrelationships of different elements of a complex system
   - work comfortably with complex systems

5. Experiment design and execution
   - test design options
   - validate and calibrate
   - compare theory (design) and observation (performance)

6. Quantitative interpretation and inference
   - ability to use statistics in data analysis
   - ability to make error estimates
   - make valid conclusions and recommendations

- recognize and resolve discrepancies
- evaluate claims of others

7. Publication and documentation of results
- recognize the requirements of particular
documents - internal vs. external, etc.
- ability to match documents to their intended
audiences

8. Knowledge of current computer techniques and software
- keep pace with rapidly evolving technology
- help co-workers become comfortable with new
technology

9. Communication, listening, teaching skills
- understand what customers want
- understand what your management wants
- facilitate understanding between groups (scientists - engineers)
- summarize, explain complex issues of concepts
- recognize marketing opportunities

10. Work as part of a research team
- respect and consider the ideas of team members
- work across discipline lines
- develop effective leadership skills

11. Appreciation for continued learning
- desire and ability to learn new skills
- move into new areas as interests and opportunities evolve
- use the library, read journals, attend symposia to
maintain awareness of new ideas and trends.

12. New scientific interests
- maintain familiarity with current and future
technologies
- understand current and future scientific priorities
- recognize future business opportunities

Self assessment is not the only means of identifying and evaluating potential jobs, but it is important. It gives you direction because it helps you understand your strengths, weaknesses, desires, and loathings. It provides you with background information you may need as you converse with prospective employers.

> **I earned my BS in physics in 1992. While still attending school, I worked as a summer intern at Boeing in their Product Development Noise Control Engineering group and I found that a combination of the love of airplanes and my degree led me to the right place. I've done various jobs, working from solving a whole host of interior noise problems before airplanes are delivered to doing airport noise assessment for a 2005 High Speed Civil Transport.**
>
> Michelle Gross, Boeing

> **This life is a test; it is only a test.**
> **If it were a real life,**
> **you would receive instructions**
> **on where to go and what to do.**
>
> Unknown

PHYSICS TRENDS

Communication and People Skills

Percentage of physics bachelors who spend a large amount of time on the following work activities, 5-7 years after earning their degrees

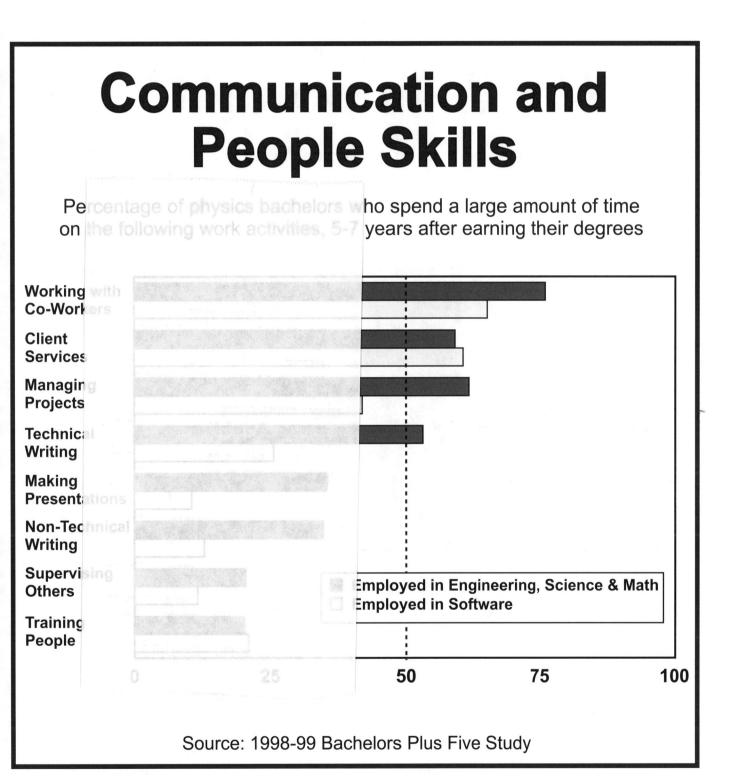

Working with Co-Workers

Client Services

Managing Projects

Technical Writing

Making Presentations

Non-Technical Writing

Supervising Others

Training People

☐ Employed in Engineering, Science & Math
☐ Employed in Software

0   25   50   75   100

Source: 1998-99 Bachelors Plus Five Study

AMERICAN INSTITUTE OF PHYSICS

Statistical Research Center
www.aip.org/statistics

PHYSICS TRENDS

What's Important?

Importance of knowledge and skills for physics bachelors, 5 - 7 years after degree

Primary Field of Employment

A bar chart comparing the percent "Very Important" for various knowledge and skills across two primary fields of employment: "Engineering, Math & Science" and "Software".

Skill	Engineering, Math & Science	Software
Scientific problem solving		
Synthesizing information		
Mathematical skills		
Physics principles		
Lab or instrumentation skills		
Scientific software		
Knowledge of physics		
Modeling or simulation		
Product design		
Computer programing		
Software development		

Percent "Very Important"

Percentage of physics bachelors who chose 4 or 5 on a 5-point scale where 5 = essential.

Source: 1998-99 Bachelors Plus Five Study, preliminary results

AMERICAN INSTITUTE OF PHYSICS

Statistical Research Center

www.aip.org/statistics

Resources for the Job Hunter

If you have spare time (!), the following pages contain the identity of various resources: sources of employment-related data, books, publications of various kinds, directories, employment guides, databases, internet sites, and corporate home pages. This is a reference section - to be used, not read! In many ways, the internet is a rich resource for the job hunter.

Employment Data: Statistical Research Center, American Institute of Physics

The American Institute of Physics's Statistical Research Center (SRC) collects data on the composition and dynamics of the scientific labor force and the education system - each with a focus on physics. Below is a list of their current publications along with a brief description of each. Unless otherwise indicated, single copies are available free of charge from www.aip.org/statistics. Also, you can call (301) 209-3070 or send your order to SRC, American Institute of Physics, One Physics Ellipse, College Park, MD 20740. (If there is a charge, enclose check payable to SRC).

Physics and Astronomy Senior Report - Looks into the backgrounds, experiences, and future plans of physics and astronomy majors at the point of graduation.

Enrollments and Degree Reports - An examination of academic enrollments and degrees conferred in physics and astronomy programs nationwide.

Graduate Student Report - A summary of the characteristics and career goals of physics and astronomy graduate students.

Initial Employment Report - A description of the initial employment and continuing education of physics and astronomy degree recipients.

The Physics Job Market: From Bear to Bull in a Decade - A discussion of supply and demand issues in physics, including historical employment trends.

Physics in the High Schools IV - An analysis and interpretation of information collected in a nationwide survey of teachers of physics at the secondary level.

Academic Workforce Report - A detailed analysis of faculty openings and new hires in universities and four-year colleges.

Roster of Physics Departments with Enrollment and Degree Data - Detailed data for physics degree-granting departments in the United States.

Salaries: Society Membership Survey - An analysis of the effect of factors such as geographic location, employment sector, gender, years from degree, and degree level or salary levels and salary increases. $15 for single copy, $10 for each multiple copy.

Who's Hiring Physics Bachelors? - A state-by-state listing of many of the companies that recently hired new physics bachelors. See www.aip.org/statistics/trends/emptrends.htm

Physics-Related Professional Societies Employment Support

Acoustical Society of America (ASA)   www.asa.aip.org
- Online career information
- Link to AIP

American Association of Physicists in Medicine (AAPM)   www.aapm.org
- Jobs in print
- Link to AIP

American Association of Physics Teachers (AAPT)   www.aapt.org
- Job service online
- Job center at meetings
- Link to AIP

American Astronomical Society (AAS)   www.aas.org
- Career services, including job registry
- Online job listings
- Link to AIP

American Crystallographic Association (ACA)   www.hwi.buffalo.edu/aca
- Online job listings
- Link to AIP

American Geophysical Union (AGU)   www.agu.org
- Online job listings

American Physical Society (APS)   www.aps.org
- Online careers in physics
- Employment Centers at meetings
- Career workshops
- Link to AIP

American Vacuum Society (AVS)   www.avs.org
    • Job center at meetings
    • Link to AIP

Optical Society of America (OSA)   www.osa.org
    • Employment centers at meetings
    • Employment website

Society of Rheology (SoR)   www.reology.org/sor
    • Link to AIP

Print References: Books and Magazines

The following list is compiled from references which can be of assistance in your search. This list is only a handful of the many existing sources for the job seeker. Since so many references exist, the American Institute of Physics has selected seven, listed at the beginning, with which we recommend you begin your job search.

- *Physics Today.* A monthly professional scientific magazine with Help Wanted ads directed to an audience of physicists, American Institute of Physics.

- *Put Your Science to Work: The Take-Charge Career Guide for Scientists*, Peter Fiske, American Geophysical Union, Washington, DC, 2001.

- *Corptech Directory of Technology Companies.* This is published annually. Coverage of over 40,000 high-tech manufacturing companies of all sizes. Corporate Technology Services, Inc., Woburn, MA.

- *The Hidden Job Market, 2000.* Peterson's Guides, Princeton, NJ 1996. A list of 2,000 firms considered to have growth potential.

- *Rethinking Science as a Career: Perceptions and Realities in the Physical Sciences,* Sheila Tobias, Daryl E. Chubin and Kevin Aylesworth, Research Group, Tucson, AZ, 1995.

- *Scientific and Engineering Societies: Resources for Career Planning*, American Association for the Advancement of Science, Washington, DC.

- *Careers in Science and Engineering: A Student Planning Guide to Grad School and Beyond*, National Academy Press, Washington, DC, 1996.

- *Scientists and Engineers for the New Millennium: Renewing the Human Resources*, sponsored by the Alfred P. Sloan Foundation, March 2001.

The final AIP recommendation is not job oriented *per se*; rather it is a book by a first-rate scientist that has interesting things to say.

- *Advice to Young Scientists*, P. B. Medawar, Basic Books, 1979.

AMERICA'S FEDERAL JOBS
Jist Works Inc., Indianapolis, IN.

CAREERS IN CHEMISTRY - QUESTIONS AND ANSWERS
American Chemical Society, Washington, DC.

CAREER OPPORTUNITY UPDATE
Monthly magazine. Career Research Systems, Santa Ana, CA.

CHAMBER OF COMMERCE DIRECTORIES
Available in many cities, and restricted to the areas of service.

CHRONICLE OF HIGHER EDUCATION
Weekly publication. Washington, DC.

CORPORATE JOBS OUTLOOK
Monthly publication profiling 20 companies per month.
Corporate Jobs Outlook, Bourne, TX.

THE DAMN GOOD RESUME GUIDE
Yana Parker, Tenspeed Press, Berkeley, CA, 1996.

DIRECTORY OF AMERICAN RESEARCH & TECHNOLOGY
Organization active in product development for business, R. R. Bowker, Division of Reed Publishing, New York.

DIRECTORY OF DIRECTORIES
Gale Research, Inc., Detroit.

DUNN AND BRADSTREET MILLION DOLLAR DIRECTORY
Dun & Bradstreet, Inc., New York.

THE ELECTRONIC RESUME REVOLUTION
Joyce Lain Kennedy and Thomas Morrow, PhD, John Wiley and Sons, 1994.

EMERGING CAREERS: NEW OCCUPATIONS FOR THE YEAR 2000 AND BEYOND
Feingold and Miller, 1983.

EMPLOYMENT GUIDE FOR ENGINEERS AND SCIENTISTS
Institute of Electrical and Electronics Engineers, Inc., New York.
Also on-line

ENCYCLOPEDIA OF ASSOCIATIONS
National Organizations of the U. S., Gale Research, Inc., Detroit.

GETTING WHAT YOU CAME FOR: THE SMART STUDENT'S GUIDE TO EARNING A MASTER'S OR PHD
Robert L. Peters, NoonDay Press, New York. 1992.
National Association of Colleges and Employers, (NACE) Bethlehem, PA, 1996.

JOB HUNTER'S SOURCE BOOK
Gale Research, Inc., Detroit.

KNOCK 'EM DEAD
Martin Yate, Adams Publishing, Holbrook, MA, 2002.

THE SCIENTIST
Biweekly newspaper for the science professional. The Scientist, Inc., Philadelphia.

STANDARD & POOR'S REGISTER OF CORPORATIONS, DIRECTORS AND EXECUTIVES
Standard & Poor, New York.

THOMAS' REGISTER OF AMERICAN MANUFACTURERS
Thomas Publishing, New York.

WHAT COLOR IS YOUR PARACHUTE?: A PRACTICAL MANUAL FOR JOB HUNTERS AND CAREER CHANGES
Richard N. Bolles, Tenspeed Press, Berkeley, CA.

Check your campus career center and their website for these and additional job-search references.

Electronic Resources

Resume Databases

Different organizations - for example, professional societies as well as commercial employment agencies - provide a service, free or for a fee, that permits individuals to post their resumes in an electronic database. These databases are accessible to employers who can search them to find potential employees. You should check to see if your campus career center, physics department, or alumni office have resume databases for organizations seeking graduates from your college/university. If you want to file your resume in a database, a different kind of resume is required.

We have stressed that a resume is targeted to a particular employer and a particular job. By contrast, a resume filed in an electronic database is NEITHER employer specific NOR job specific. It is a generic resume studded with key words that might be used by employers for searches. What first-time job seeker should consider putting her/his resume on file? The prime candidates are those who have specific knowledge and/or skills that are likely to be the object of a search. For example, job seekers with a PhD whose research has been conducted in a hot area could consider posting their resumes.

A database resume is generic.

It is not targeted.

The following are resume databases:

American Institute of Physics
Career Services Division
CAREER OPPORTUNITIES
(301) 209-3190
http://www.aip.org/careersvc/

BRASSRING.COM
http://brassring.com/

FLIPDOG.COM
http://www.flipdog.com/

HOT JOBS.COM
http://hotjobs.com/

MONSTER.COM
http://www.monsterboard.com/

PHOTONICSJOBS.COM
http://www.photonicsjobs.com/

THE PHOTONICS EMPLOYMENT
http://www.photonics.com/employment/

POST-DOCS.COM
http://www.post-docs.com/

SCIENCE MAGAZINE.COM
http://www.recruit.sciencemag.com/jobsearch.dtl

The web is a powerful resource for the job hunter.

It is no substitute, however, for a network

of well-placed individuals who can help you.

Web Sites with Job Listings

The internet provides many sites with job listings. The internet is fast and links the user to jobs throughout the world. Here are sites to consider.

Physics web sites:

http://www.aip.org/careersvc/ The American Institute of Physics Career Services Division. Offers job postings, information on services provided by the division, and information pertaining to careers in physics.

http://www.aip.org/ptbg/search.jsp The *Physics Today Buyer's Guide* with a searchable database of corporations. 1,991 corporations are listed with location and product line.

http://www.hep.net/employment/indiv-jobs.html High Energy Physics Information Center job postings.

http://tiptop.iop.org/ Institute of Physics (IOP) job postings.

http://physicsweb.org/ Institute of Physics (IOP). Offers physics news, international jobs and resources.

http://www.psrc-online.org/ Physical Sciences Resource Center. Search Online Classifieds Database using keyword.

Optics and Laser sites:

http://optics.org/jobs/ Online Photonics Resource.

http://www.workinoptics.com Built by the Optical Society of America.

http://www.photonics.com/employment/ Photonics Employment Center. View jobs in 9 different categories.

http://www.opticsjobs.com/Companies/joblinks.htm Optics jobs listed by company.

Engineering web sites:

http://www.engineerjobs.com/   Engineer jobs, the engineering job source, provides fast-loading and easily accessible job listings for engineers and technical professionals.

http://ieee.org/ The IEEE home page contains an extensive job listing in a diverse range of electronics-based industries.

http://www.engcen.com/ Listing jobs for engineers since 1997.

Science Web Sites

http://www.nature.com/naturejobs/ Jobs in all scientific categories.

http://recruit.sciencemag.org/jobsearch.dtl Science magazine job search in life sciences and other sciences fields.

http://www.the-scientist.com/   Job listings for The News Journal for the Life Scientist.

http://www.phds.org/ PhDs.org Science, Math, and Engineering Career Resources.

http://www.post-docs.com/ Source for post-doctoral opportunities.

http://chronicle.com/jobs/ The Chronicle of Higher Education magazine career network. Browse jobs in academia.

http://www.scijobs.org/ Indexes in a wide range of science related jobs.

Computing web sites:

http://www.cra.org/main/cra.jobs.html The Computing Research Association's jobs web page. Positions for Computer Scientists, Computer Engineers, and Computer Researchers

http://www.acm.org/cacm/careeropps/ Association for Computing Machinery Career Opportunities.

http://www.computerjobs.com/homepage.asp Thousands of computer jobs updated hourly.

National Job Boards:

http://www.Headhunter.net/JobSeeker/Index.htm?siteid=cmhome Searches database of jobs nationwide.

http://vertical.worklife.com/onlines/careermag/ Career Magazine. Search jobs by profession.

http://www.monsterboard.com/ Monster.com job board. Search by location and job category.

http://www.dice.com/jobsearch/ Information Technology (IT) job board with high tech permanent, contract and consulting jobs.

Employment Pages at Government Labs

Los Alamos National Laboratory
http://www.hr.lanl.gov/

National Institute of Standards and Technology (NIST)
http://www.nist.gov/public_affairs/employment.htm

Fermilab
http://fnalpubs.fnal.gov/employ/jobs.html

NASA
http://www.nasajobs.nasa.gov/

National Renewable Energy Laboratory
http://www.nrel.gov/hr/employment/

Oak Ridge National Laboratory
http://ornl7.ornl.gov/employment/

Brookhaven National Laboratory
http://www.bnl.gov/HR/jobs/default.htm

Argonne National Laboratory
http://www.hr.anl.gov/employment2.htm

National Energy Research Scientific Computing Center (ERSC)
http://www.lbl.gov/CS/Careers/OpenPositions/

Corporate Home Pages

Most corporations have job information linked to their home pages. By examining these you can obtain a sense of the type of positions that are open. To help you get started, there follow two lists of companies compiled by the Statistical Research Center of the American Institute of Physics. The first list gives the companies that employ the most PhD physicists. This list certainly identifies companies where the work done draws from physics. The second list gives the companies, broken out by state, that have recently hired baccalaureates in physics.

The Corporations That Employ the Most PhD Physicists

These corporations employ nearly one third of all physicists in the private sector and are listed in descending order of the number of physicists they employed over the period 1996-2000. Since these data were collected, many of these corporations have undergone major restructuring, including mergers, divestitures and downsizing. The order of the top 9 companies may change but the top 9 will remain in the top 9. Also company names may have changed, e.g. AT&T became Lucent Industries, and Lockheed Aircraft Corporation became Lockheed Martin. Data are based on a sample survey from all 10 AlP Member Societies. The information is from the AlP's Statistical Resource Center.

Lucent Technologies
IBM
Lockheed Martin
Science Applications International Corporation
Boeing
Raytheon
General Atomics
Schlumberger
Eastman Kodak
(The corporations listed above employ one fifth of all the physicists in the private sector)

Hewlett-Packard Company
Northrup Grumman Corporation
Motorola Incorporated
Rockwell International Corporation
Maxwell Technologies
Varian Associates
3M Company

**Step 4
Make appointments
at two or three companies
for informational interviews.**

PHYSICS TRENDS

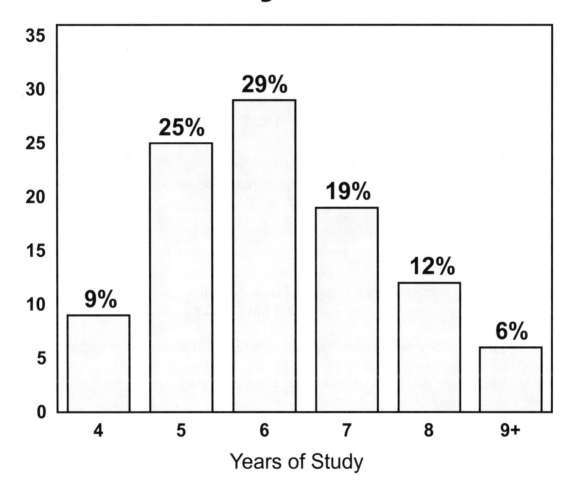

How Long Does It Take to Get a Physics PhD?

9%
25%
29%
19%
12%
6%

4   5   6   7   8   9+

Years of Study

This graph depicts the number of full time equivalent years of graduate study completed by the PhD class of 2000.

Source: Initial Employment Report

AMERICAN INSTITUTE OF PHYSICS

Statistical Research Center
www.aip.org/statistics

NETWORKING II

You move to a new metropolitan area. You hardly know anyone: a couple of neighbors, a few people where you work. You need a physician. What do you do? Pick a name out of the yellow pages? Not likely. Most likely, you ask your neighbor or a friend at work: Who is your doctor? What doctor would you recommend? That is networking. Or you need a job. You talk to people who are "close to the job you want." That is networking.

> **Having a good education and good test scores are very important, but it DOES matter if you know someone close to the job you want.**
>
> Larry Collette, General Services
> Administration,
> Ft. Worth

There is nothing strange about networking. It goes on all the time, in all locations, by all people. It goes on in very casual ways and in formal ways; for example,

- You ask a classmate about Professor Newton. Is she a good teacher?

- You ask the cashier, a total stranger, at your gas station to suggest a good auto mechanic.

- You tell a friend about your favorite pizza spot.

- You ask your roommate to find out where her sister got her shoes.

All this is networking. In fact, almost every person you meet and every conversation you have is a potential form of networking. As your career gets started, you will naturally develop a network of individuals you can call upon for assistance and advice and, at the same time, you will be in the network of others and they will call upon you for assistance and advice.

It is never too early to start networking. Right now, while you are a student, you can begin to build a network that will prove very helpful when you begin actively job hunting.   The first step is the hardest; however, once started, you will be amazed how quickly your network of contacts grows.

> ## It's not easy to ask a stranger for advice,
> ## but
> ## it's not hard either.

As a student, what is the purpose of building a network? And how do you begin?

Since the immediate purpose of building a network of contacts depends on how long before you actually start looking for a job, let's think in terms of two phases.

Phase I: Two years before you hit the pavement
looking for a job.

Phase II: One year before the search begins.

> ## Networking - Phase I
>
> ## Get to know physics advanced students.
>
> ## Get to know alumni.

The purpose of Phase I is, so to speak, to determine the lay of the land. As stated earlier, the one great place to start is the alumni of your physics department. You have a natural link to former physics majors and essentially all of them are pursuing interesting and fruitful careers. And most of all, these alumni would be happy to tell you about their work and to help you when your job search begins. Since advanced students are a year or two away from being alumni themselves, get to know them. Here are some specific suggestions:

- Find out where the alumni are and store this information - it can be very useful later.

- Pick out a couple of alumni members and write them a letter. Introduce yourself and tell them you would like to know about their work.

- Organize a group of students and invite some nearby members of the alumni to come and talk to your group. Set up a series of talks.

It may be a challenge to get information about the physics alumni. Ask a physics professor or someone in the department office for the names and addresses of the physics alumni. If such a list does not exist or is out of date, try the campus alumni office; people there may have more complete and accurate alumni information. Many campus career centers maintain active networks of alumni and other individuals who have indicated a specific willingness to meet and/or talk with students.

A third source is your immediate and extended family. Aunts, uncles, or cousins may work for companies that employ engineers and physicists. A family member provides a natural channel to strangers. Finally, don't forget family friends and contacts: neighbors, the family physician, friends at your place of worship, etc.

**At every step along the way,
always thank people who have helped you.
Don't forget.**

The outcome of Phase I is the names, addresses, and telephone numbers of individuals you can call upon with specific questions, for advice, and for help.

**Step 5
Target two or three potential employers and
study them in depth to learn what they do
and how they generate their revenue.**

In Phase II, your networking gets more refined. By this time you are closer to graduation and may have a better idea of the type of job you want and perhaps even the employer you would like to work for. Pick out the most appropriate contacts you made in Phase I and contact an appropriate individual by letter, phone, or e-mail. You can get good advice from your campus career center, regarding appropriate and inappropriate questions to ask. To begin, address the individual respectfully by name. If possible, make an appointment for a visit. Identify and discuss your interests; identify your talents and abilities; let this individual know you are a willing worker. Inform him/her how your thinking about employment has developed. Seek his/her judgment about options you are considering. Be as specific as possible. Ask for recommendations. Get the names of individuals you can contact who are employed by those companies that interest you. Through it all, let yourself shine.

Networking - Phase II

Expand your network to individuals who connect directly or indirectly to employers of interest.

The outcome of Phase II is the names of individuals you will contact when your job search actually begins.

When your job hunt actually begins, your efforts in Phases I and II can pay off enormously. You are going to apply for a job at a particular company and perhaps for a particular position at that company. Write, call, or visit the most appropriate individual you have talked with and ask for their guidance. Ask him/her to critique your resume to see if it can be more tightly connected to the position you seek. Also, ask her/him who you should contact within your potential employer and ask if you could use his/her name as a reference.

Networking is important. It is the most successful means to a job. Studies show that over two-thirds of all open positions are filled through personal contact, that is, networking. The other third, sometimes called the published job market, is filled in response to published "help wanted" ads, agencies, or direct contact. Act immediately if and when the opportunity you desire presents itself. Doors can close quickly, and any perceived hesitancy on your part may cost you a golden opportunity.

Networking is not a selfish pursuit. If you are willing to network for yourself and others, then others may be willing to do the same for you. This is networking in its broadest sense.

> **Graduating seniors make contacts and get jobs at the companies represented by the attendees at our annual alumni dinner. The focus of the dinner is "network and use your contacts." We can cite that this really works.**
>
> Chris Gould, Department of Physics
> North Carolina State University

Tips For Networking

1. **Maintain a card file or spreadsheet of your contacts.** When you meet someone new, write down his or her name, job title, company address, and phone/fax number, and e-mail address. It helps to make a few notes for each contact for future reference: who the person is, the circumstances in which you met, what next steps you may take next, etc.

> **In the steps ahead, you must be able to address individuals by name.**
>
> **No "Dear Sir/Madam" letters.**

2. **Join a professional society.** The American Institute of Physics has ten member societies that are all physics related. As a member of one of these societies, you will receive *Physics Today* each month. (See the AIP web page www.aip.org). If possible, attend a national meeting of the society you have joined. These conferences are held in different parts of the country and provide an excellent opportunity to meet people who can be part of your network.

3. **Request to be put on mailing lists.** Add your name to mailing lists for organizations associated with industries that interest you. By receiving their various newsletters and publications, you can keep abreast of the conferences and events where you might meet potential contacts. More important, you will begin to link the names of individuals with activities that interest you.

4. **Attend local lectures and paper presentations.** Make it a point to introduce yourself to the speaker and comment on his or her talk. If you have published in the same area, offer to send your work to the person. Be sure to add his or her name to your card computer file. You should also introduce yourself to those sitting near you at the session. Try to find out where they are employed and, if appropriate, tell them you are interested in finding a job.

5. **Read appropriate magazines/journals.** One magazine to look at is *The Industrial Physicist*. This magazine reveals things that are going on in various industries. You will learn where the authors work and, as their e-mail addresses are typically given, you can contact them.

As a senior physics student, I find *The Industrial Physicist* to be a useful and interesting magazine. I think more students should be exposed to *TIP* to help further their understanding of the industrial physics workplace....

Mark Lentz, Northwestern State University

6. **Follow up on all leads.** If a contact tells you to get in touch with him or her or with someone he or she knows, make sure you call or write promptly. Be sure to send a thank you note back to the person who gave you the lead.

7. **Contact people directly.** Even if you are shy, do it. By appointment, your campus career center will often provide a "practice session" to help you prepare. Always address people by name. You will find that people are willing to help.

> **Everyone knows people
> who know people
> who know people
> who know people...**
> *ad infinitum*.

The "Nifty Fifty"

Everyone has a network. You know people who know people. Identify these individuals. It is not so tough. Everyone you know is a potential contact. Use a template such as the one below to start identifying your first fifty contacts.

- Immediate family members.

- Extended family members (cousins, aunts, uncles, etc.).

- Close friends.

- Extended friends (friends of friends, neighbors, everyone in your address book.).

- Everyone at the business places and retailers that you visit.

- Everyone in your leisure activity circles.

- Every doctor, medical professional, or other professional you know.

- Every professor, teacher, lecturer, or demonstrator you know, once knew or have worked with.

- Every clergy member you know.

- Every person in your place of religious activity.

- Everyone you come in contact with or are introduced to...EVERYWHERE Grocery stores, gas stations, postal workers, repair persons...EVERYONE!

The lesson of the "nifty-fifty" is this:

People you know well are the means to build a robust network of valuable contacts.

Action Points II

_____1. Develop a network of contacts both inside and outside your discipline and both on and off campus to help you understand the full range of opportunities available to you.

_____2. Constantly review advertisements in science and physics magazines, journals and newspapers in cities where science and physics employment centers are.

_____3. Know yourself so that when you see an opening, you can market yourself, your education, skills, and attributes.

_____4. Visit your university's career center or library to read some of the many publications that provide guidance in resume preparation and interviewing.

Action taken early

will avoid frustration later.

PHYSICS TRENDS

Considering a Career in High School Teaching?

Teacher certification requirements are different in each state and many states offer several paths to certification. Check out:

www.aip.org/statistics/trends/hstrends.htm

to find out whom to contact to get the latest information on certification requirements in your state.

AMERICAN INSTITUTE OF PHYSICS

Statistical Research Center
www.aip.org/statistics

THE SEARCH

> **The people who get hired are not necessarily those who will do the job best, but those who know the most about how to get hired.**
>
> Richard N. Bolles

"The Search" means getting into the nitty-gritty of determining

- the sectors where the jobs are most abundant,

- which sector fits your interests,

- which job(s) within those sectors you are going to concentrate on and learn about, and

- which particular jobs you will eventually apply for.

For a physicist, this means knowing how to look for and interpret job requirements that are appropriate for you as a physicist. **Remember:** Rarely are open positions announced "for physicists." **Also remember:** As a physicist you are qualified for many positions whether or not they are identified "for physicists." (Notice the job titles in the box below.)

My jobs, and their actual titles, have included software developer, manufacturing engineer, mechanical-design engineer, reliability engineer, project manager, and director of quality assurance. None of them has given me more satisfaction than working in quality engineering, which offers challenging problems and is intellectually and financially rewarding.

Mark Annett, Cerebral Palsy

Getting Started

Step 6
Expand your network to include individuals who are employed by your targeted employers. Talk to members of your network about these employers.

To Get a Job:

- **Know yourself.**
- **Know your would-be employer.**
- **Talk to someone working for your would-be employer.**

Know what it means to be a physicist

As a first-time job seeker, what does it mean to be a physicist? It means you have a broad and powerful technical foundation. It means you can do many of the things, perhaps most of the things, that an engineer can do. It is important for you to recognize this **because** employers in the technical arena do not label their positions "Physicist." They label them Engineer or Computer Something-or-Other, but rarely Physicist. So as a physicist you must, more-or-less, ignore the job title; instead, read the job description, understand the job requirements, and determine whether you can meet those requirements. Your education has prepared you to meet successfully the job requirements of most positions. Know that you can structure your resume and interview to bring out the match between the needs of the job and your qualifications.

Target jobs that match your ability

You have done some kind of self assessment. You know yourself. Match your abilities with the requirements of the job. To some employers, it can be just as bad to be over qualified as it is to be under qualified. Know your capabilities and the level at which you can operate. If you are under qualified, the employer will not seriously consider you. On the other hand, if you are over qualified, employers are usually reluctant to hire you for fear that you will become bored or soon receive a better offer.

> **Play to your strengths.**

Know your strengths. Also, know what is in demand. Read help-wanted ads to learn what employers are seeking, those skills that are in demand, which industries are hiring, and what "buzz words" are in common usage. This knowledge can help you link your strengths to an employer's needs.

Adopt a search technique that plays to your strengths

Take advantage of your natural talents when deciding how best to approach employers. If you have a particular talent for writing, use it for writing creative resumes and cover letters. Develop a portfolio of your writing and indicate its existence in your resume or cover letter. If research or problem solving is your particular strength, go to the library and investigate potential employers' points of interest. What kinds of problems are of commercial interest to them? The most important thing is to put your strengths on display. This will not only enhance your comfort level, but it will make you more appealing to employers as you present your ideas and strengths with conviction.

Be prepared to spend time on your search

Finding a job can take time. It can be frustrating. Be prepared. As stated above, advance preparation can minimize the frustration.

> **The average job search takes 6-9 months, so think of your job search as the hardest three-credit class you have taken. That is about how much time you should be spending each week researching organizations and jobs, crafting application materials that are specifically targeted to each opportunity, and continuing to build and use your network.**
>
> Linda K. Gast, Director, Career Center, University of Maryland

> **The better your network,
> the easier the ordeal.**

Be aware of the "blind-ad" strategy

Employers often advertise in the classified section without giving their names or numbers. The reasons for this vary. The ad may have been placed by a headhunter advertising on behalf of a client, or the employer may want to review a wide range of resumes. The chances of getting a job (or even a response) from a blind ad are slim; however, such ads should not be dismissed. The point is, don't make this your primary strategy.

Don't overlook small businesses

According to a study by Dunn & Bradstreet, 80 percent of all new jobs are being created by companies with 100 employees or fewer. Small businesses offer a large potential for growth, can often provide greater levels of responsibility, and provide greater opportunities for advancement. When targeting small businesses, you should use strong networking strategies and display assertiveness. Often the hiring manager and the president are one and the same, and he or she is likely to appreciate an entrepreneurial approach (see below, *The Hidden Job Market,* by Peterson).

Number of Companies Performing R&D
(by number of employees)

Fewer than 500	35,112
500-999	1,127
1,000- 4,999	1,302
5,000-9,999	332
10,000 - 24,999	199
Over 25,000	167

National Science Foundation, Science and Engineering Indicators, 2000, p. 2-24.

It was a big step for me to leave academia and Physics and try to find a job in Engineering. I started by perusing the want ads. Most of the ads went something like this, "Entry level engineer, BSEE/MSEE req. Must have experience or training in....." and there would follow a string of three letter acronyms I not only had no training in, but had never heard of. After a few months of desultorily looking at these ads, I decided that I needed help. So I went to my campus career center.

I cannot recommend this highly enough. I found exactly the help I needed there. I had weekly appointments with a career counselor, who helped me write a good resume and cover letter. (Incidentally, she would have helped me define what I wanted in a job if I had not been very clear about that already.) She did not specialize in technical fields, and so was not helpful in finding places to apply to, but as it turned out that was unnecessary.

So, one day in my occasional reading of want ads, I came across an entry level RF engineering position, where they wanted a BS or MS in EE, Physics, Math, Chemistry, Computer Science, etc. I went to the company's web site and was impressed with the corporate culture presented there. Also, the job sounded interesting, and touched on two topics (superconductivity and microwaves) that I had worked with as an undergraduate research assistant. So, I sat down, updated my cover letter for that particular company and faxed it off immediately with my well-polished resume. And the resume did its job – after months of worry, that turned out to be the only resume I sent. After the interview I was sure it was the right job for me. In my three years there I never doubted my decision.

Abigail Davis, Engineering Physicist, EVI

Step 5 – Again
Target two or three potential employers and
study them in depth.
Know what they do and
how they generate their revenue.

Step 6 – Again
Expand your network to include
employees, or people who know employees,
of your targeted companies.

Talk to these individuals, quiz them
about the target companies.

I begin with a story, a true story.

A True Story

It occurred in San Jose, California. In silicon valley country. It was a meeting of the American Physical Society and I had arranged a session with a Hewlett Packard physicist as the speaker. The room was packed with job-seeking physicists.

The first image that appeared on the screen was an organization chart of Hewlett Packard. "How many of you have seen an organization chart," asked the HP scientist. Some had seen one, most had not. "I thought so," continued the speaker. The box at the top of the chart identified the CEO of HP and a row of some eight or ten boxes at the bottom identified the different divisions within HP. "I am showing you this," and his voice lowered to almost a whisper as he ran his finger along the row of divisional boxes, "because these boxes identify how we make our money. If you are going to work for HP," the speaker continued, "you must show us how you are going to fit into one of these boxes and help us make money."

John S. Rigden
American Institute of Physics

What is an organization chart? It looks something like the figure below. Of course, the number of levels as well as the number of boxes on each level varies from corporation to corporation. Also, each box has a label that identifies what it represents.

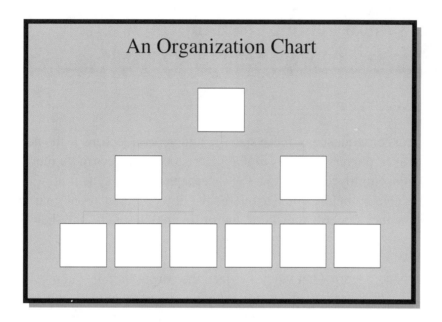

Of course the Hewlett Packard physicist said much more than appears in the box above, but he was conveying an important message to the physicists in the audience. What was his message? It was this: If you are going to apply for a job at Hewlett Packard, you must know what is in those boxes along the bottom of the organization chart. You must know what each division does with sufficient specificity that you can identify how you fit into a particular division's mission, into a particular division's day-to-day activities. In short, you must recognize how you can contribute to the mission and enhance the revenues of HP.

The message in the above story transcends Hewlett Packard. The general lesson is that you must learn about the company before you submit a job application. Your knowledge about the company must then be integrated into your job application and into your job interview. To begin with, you **must** know the name of the person to whom you are submitting your job application. A "Dear Sir/Madam" letter will not do the job. This is why a form letter is a waste of time. The individuals at corporations who look at job applications receive hundreds upon hundreds of applications and something has to capture their attention immediately or it is placed on the wrong pile - the reject pile.

> **Never, never send a form letter
> to a would-be employer.
> Such a letter shows that you are shooting
> in the dark and
> know nothing about the employer.
> It is a waste of time and postage.**

The practical consequence of these specific and general lessons is the need to focus on a couple of potential employers. It takes time to learn about a corporation; therefore, you must narrow your potential employers down to a few - perhaps three, maybe four. Of course, you begin with a larger number before you narrow the list down. Learn about your choice companies in some detail: What they do, how they do it, their principal products, and their prime customers.

Also, you want to learn what kind of citizen the corporation is. You may, you may not, have some things you feel strongly about; for example, diversity or environmental issues. You do not want to discover, after you start working, that your employer is doing things that make you very uncomfortable. You must be careful how you go about gathering such information as these can be touchy issues.

> **Learn about the employer.**
>
> **Do you want to work there?**

> **Learn about the employer.**
>
> **Import that knowledge into**
>
> **your cover letter and resume.**

How do you learn about a company? There are several ways. Here are some of them.

Go to the corporation's web page

A good place to start. Many corporate web sites have job information as a part of their home page.

Write and ask for the annual report

The annual report can often be downloaded from the corporate web site.

Use your network

Talk to individuals about the companies you want to work for. Ask them if they know any one who works there.

Schedule an informational interview

Request an informational interview. Here, once again, is where your network comes to the fore. Find out from your contacts whom in the organization you should seek to interview. Contact this person by e-mail or telephone and request an informational interview so that you could learn more about the company. Tell him/her you are interested in possible employment.

An informational interview may give you the opportunity to talk to an insider who may be doing work similar to what you are interested in. The purpose here is to find out about a company and types of jobs before actually applying. Make the interview short, no more than 30 minutes (15-20 minutes more typical).

> **A successful informational interview**
>
> **often leads to a job offer.**

The informational interview is different from a job interview. With the former, you are in charge; with the latter, the interviewer is in charge. In an informational interview you can ask questions that would be inappropriate in a job interview. For example, you can ask: How did you get interested in this position? How did you get hired? What excites you most? What do you like least about your job? What kinds of challenges do you deal with? What skills do you need? What is the company like to work for?

At the end of your interview, be sure to thank the person for taking time to meet with you. Also be sure to write a thank you note later that day. This sort of interview is also a good way to practice your interviewing skills.

Arrange an internship

It is possible that you could get an internship after graduation and spend time inside the corporation getting to know the ins and outs of the company. This could be a way of sampling a couple of different companies. You would most likely be assigned to work with a particular individual who would become a mentor. Internships are common in research and government agencies. Refer to Peterson's *Internships,* 1996.

Step 7
**Through conversations
with members of your network,
by examination of
your potential employer's web site,
or through help-wanted ads,
determine the type of position
you are going to apply for.**

PHYSICS TRENDS

Where Do Physics Masters Go?

US Citizens Only

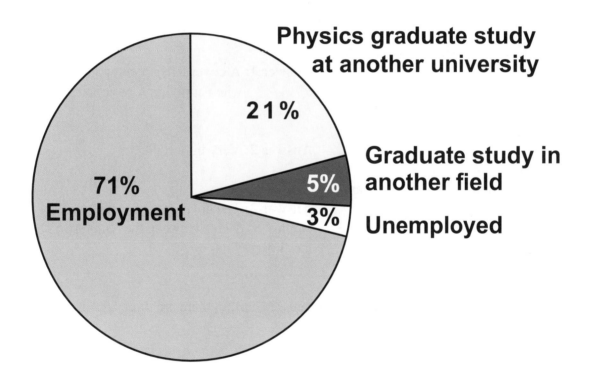

Physics graduate study
at another university

21%

Graduate study in
another field

5%

3%

Unemployed

**71%
Employment**

Career paths of physics masters 6 months after receiving their degrees

Source: Initial Employment Survey, Classes of 1999 & 2000

Statistical Research Center

www.aip.org/statistics

COVER LETTERS

> **A cover letter
> either
> stimulates interest
> or
> squashes it.**

Question 1: What is a cover letter?

> **Answer 1:** A cover letter accompanies a resume when applying for a job.

Question 2: Is it important?

> **Answer 2:** Very important.

Question 3: Why is it important?

> **Answer 3:** The cover letter is often the way employers decide between the "Interesting Pile" and the "Forget Pile."

In addition to the background work discussed earlier, there are three specific steps to landing a job:

- first, writing a cover letter,

- second, creating a resume, and

- third, getting an interview.

The purpose of a cover letter is to get its reader to examine your resume. It is a necessary part of a job application and it must accompany the resume you send to a potential employer.

Format of a Cover Letter

> **Do not call attention
> to yourself
> by an odd-ball letter format
> or
> by gee-whiz, slang, and flip words.**

The cover letter is a one-page business letter and should follow the standard format of a business letter. It should be neatly printed and centered on letter size (8 ½" by 11") stationery. Send a clean copy - no printer smudges, coffee stains, or xeroxed copies. Keep a copy of your cover letter so you have a record of what you wrote, the date the letter was sent, and the name and title of the person to whom it was sent. Writing a knowledgeable and error-free cover letter is the only way to introduce yourself to a potential employer.

The elements of a business letter are the following:

YOUR RETURN ADDRESS

DATE

EMPLOYER'S NAME AND ADDRESS

SALUTATION

BODY OF LETTER

CLOSING SIGNATURE
telephone number
e-mail address

For the Baccalaureate:	For the Masters/Doctorate:
For the Baccalaureate: **A cover letter is one page. No longer.**	**For the Masters/Doctorate:** **A cover letter may exceed one page, but not two.**

Purposes of Cover Letter

The cover letter serves five important purposes:

- to make a connection with your potential employer,

- to let the employer know the position you are applying for,

- to convey, implicitly, why the employer should hire you for the position,

- to call attention to pertinent parts of your resume, and

- to provide the employer with a sample of your writing skills.

Connect with Your Potential Employer

> **You have 7 seconds to catch my attention. If you don't, your cover letter and resume go on a pile I will probably never look through again.**
>
> An HR Director speaking to a class of physics students

How do you catch the attention of the person reading your cover letter? To get their attention, you must connect immediately. To begin, you must, *must* know the name and title of the person who will read your cover letter. Address the letter to that person with the proper title. No "Dear Madam/Sir" or "To Whom It May Concern" salutation. This is the first connection. If you have met with this person, perhaps in an informational interview, the first sentence after the salutation can remind the reader of your earlier meeting. Another connection. If you have been recommended by someone known either to the person reading your letter or to another person employed by the company, identify who recommended you. A connection. Next, you make connection by demonstrating your knowledge about the employer: perhaps the name of someone you have talked to who works there, perhaps by identifying a specific process that both interests you and is an important part of the employer's business. (Here is where the results of the earlier research you did on employers come into play.) Show that you know what the company does and why you are interested in it. Still another connection.

Identify the position you are applying for

If there is a specific position you are applying for, identify the position. Show that you understand how the particular position relates to the larger mission of the company.

Sell yourself - show why you are the person for the job

In as specific terms as possible, show how your knowledge, skills, background experiences, and interests have prepared you for the particular position.

Call attention to relevant parts of your resume

If there are experiences, skills, or knowledge cited in your resume that tie you to the position you seek, you can call attention to these. Be careful, however, not to state things exactly as they appear in your resume.

Demonstrate your ability to write well

A cover letter is a serious letter so don't be flip or cute. Do not use slang. To write, "I am the perfect gent/gal for Addison Semiconductor" or "It'll be a great day for you if you hire me" will not open any doors for you.

At the same time, a serious letter need not be a dull letter. Use active verbs. Always use active voice. Accurate grammar (subject-verb agreement), correct spelling, and flawless proof-reading (no typos) are all critical. One grammatical error, one misspelled word, or one typo can leave an impression of carelessness and, therefore, be deciding...against your candidacy for the job.

> **A cover letter is not a form letter.**
>
> **Each cover letter is specifically crafted
> for a particular employer and
> for a particular job.**

A cover letter provides employers with a quick overview of your qualifications, abilities, and personality. It is therefore important to spark an employer's interest by writing a superb cover letter to accompany your resume. The first couple sentences must capture attention. Work and rework your cover letter until superfluous words are eliminated. Every word must do a job, every word must count. A good rule of thumb: Do not exceed one page.

> **Employers often make their first cut**
>
> **based on the cover letter.**

Tips on Writing a Cover Letter

1. Personalize each cover letter.

2. Demonstrate that you know about the company or organization offering the position you are applying for.

3. Make note of any previous contact with the employer.

4. If you know a person who is also known to employees at your would-be employer, identify this person.

5. Identify the job you are applying for.

6. If you are responding to an advertisement, link your experiences and skills to the job qualifications listed in the ad.

7. Sell yourself.

8. Include a sentence stating your follow-up plan. Express your willingness to meet with appropriate employees. State when you will follow up by telephone. Be sure to provide your phone number and e-mail under your signature.

> **We look for evidence that the candidate likes to think radically and broadly and can follow through in depth. The applicant should show evidence of excitement for the research work done and for other areas of his/her education. Much of this is elicited from interviews; however, the candidate should strive to give indications of this in the resume.**
>
> David Biegelsen, Xerox

Sample Cover Letters

Earlier, you were advised to take steps that would attract attention to you as a job applicant. If you followed that advice, here is where it pays off.

Following are some example cover letters. Tom was a very good student with a GPA of just under 4.0; however, he apparently spent all his time studying. He took no electives of consequence. Nicole was also a good student, but as her cover letter shows, she did much more than get good grades. Compare the cover letters of Tom and Nicole:

- Tom does not know whether he is writing to a man or a woman; Nicole knows she is writing to Mr. Patterson because she has met him.

- Note typo (typographical error) in Tom's letter!

- While grades are important, they are rarely the deciding factor. Many job applicants have commendable grades. Tom says nothing in his first paragraph except to announce his GPA. Nicole links her academic experience with a particular division of the company and identifies a part of the work they do.

- Nicole has had experience in industry and she connects her experiences, both as a student and as an intern, directly with Bingo's interests; Nicole demonstrates her recognition that communication of technical subject matter is a challenge and that she has taken steps to communicate effectively.

Don't go around saying the world owes you a living; it owes you nothing; it was here first.

Mark Twain

Some see the glass as half-empty, some see the glass as half-full. I see the glass as too big.

George Carlin

Tom's Cover Letter - Not good

Thomas Whipple
1281 Conway Road
Chesterfield, MO 63141

May 24, 2002

Human Resources
Bingo Technologies
462 Front Street
Kansas City, MO 64108

Dear Madam/Sir:

I am interested in the job opportunities at Bingo Technologies. I recently graduated with a major in physics from Washington University in St. Louis. I have had the full complement of courses required of a physics major and have a 3.94 GPA.

In my laboratory courses I have used various instruments to measure physical parameters. Often the output of the laboratory instruments went through an A/D converter so I have had some experience with computers. Through my experience I became interested in instrumentation and would like direct my knowledge and skills to the design and improvement of physical instruments.

I enclose my resume. I believe my knowledge and experience equip me to make a contribution to Bingo. I look forward to hearing form you.

Sincerely,

Thomas Whipple

Enclosure

Nicole's Cover Letter - Good

Nicole Jackson
125 Pearl Street
Laguna Beach, CA 92651
May 24, 2002

Mr. Marvin Patterson
Section Manager, Electric Measurement Division
Bingo Technologies
462 Front Street
Kansas City, MO 64108

Dear Mr. Patterson:

I enjoyed talking with you earlier this month about the work in your section. The problems you described concerning the design and development of test instruments in the Electrical Measurement Division were intriguing. I have used several Hewlett-Packard (now Agilent) instruments very similar to those you described. I would like to work in your group and I believe I could make a contribution.

I have just graduated from the University of Colorado with a Bachelors of Science in physics. In addition to the requirements of my major, I have taken two writing courses: scientific writing and creative writing. I enjoy the challenge of communicating technical topics to non-technical audiences.

As an undergraduate I worked as an intern at a small high-tech company, TLCC, Inc., for two summers where I worked closely with product engineers developing rf detectors. My responsibilities included testing and calibrating the detectors. I enjoyed my internship and I see parallels between my experiences at TLCC and the activities of your division. I believe my education, my work experiences, and my interests in instrumentation have prepared me to make a contribution to the Electric Measurement Division and to Bingo.

My resume is enclosed in which my internship is described in a little more detail. I would like to meet with you again to discuss the overlap between my background and your needs. I will contact you by phone later next week to discuss the possibility of an interview.

Sincerely,

Nicole Jackson
(440) 354 3285
njackson@bnl.com

Enclosure

March 30, 1996
44 Spruce Street
Bethesda, MD 20817

Mr. John G. Smith
President
ABC Med-Instrument Corporation
9 Main Street
Gaithersburg, MD 20879

Dear Mr. Smith:

At the April meeting of the American Association of Physicists in Medicine, I met John Sallucci who works for you. He suggested I write to you about the medical physicist position at ABC Med-Instrument Corporation.

As a recent MS degree recipient, I have spent the past year studying radiotherapy in cancer treatment at the Washington, DC Medical Hospital. I also spent several summers interning at the Columbia Hospital for Women, in Washington, DC. I have learned about the diagnostic tools produced by ABC and I believe my background in physics and medical physics meets the qualifications for the position and equips me to contribute to your product line.

Enclosed is my resume to provide additional information about my undergraduate and graduate work. I would appreciate the opportunity to meet with you to discuss further how my qualifications meet the needs of your organization. I look forward to hearing from you within the upcoming weeks. Thank you for your time and consideration.

Sincerely,

Susan P. Cho
(301) 209 4689
spc@uchi.edu

Enclosure

113 S. Childress Lane
El Monte, CA 91882

November 3, 1996

Ms. Amy Ellison
Manager Technical Staff
Texas Instruments
5872 North Season Avenue
Dallas, TX 75225

Dear Ms. Ellison:

At a Society of Women Engineers conference I attended recently at the University of California, Berkeley, I spoke with Brian Karfonta, a Texas Instruments recruiter. Mr. Karfonta suggested I send my resume and completed course list to you as I wish to be considered for an internship with TI.

As a student completing the electronics option in physics, I am interested in product development in TI's divisions engaged in developing electronic test instruments or computer systems. I am particularly interested in your high-speed amplifiers operating at frequencies greater than 50 MHz.

My work in laboratory courses and at the Computer Center at California State Polytechnic University at Pomona has provided me with valuable practical experience in electronics.

I will contact you in the next week to discuss my qualifications with you in more detail.

Sincerely,

Stephanie P. Jones
(617) 495 3460
stepjones@erols.com

Enclosure

THE RESUME and the Curriculum Vitae

A resume is not a CV.

A resume is different from a curriculum vitae (CV). The resume is used for most job applications. Government positions, which many physicists apply for, can have special guidelines. A CV is a complete record of your educational and professional life - everything but the kitchen sink. For example, in a CV you would include educational background, all positions held, a list of all publications, grants obtained, invited talks given, honors received, and all background experiences. The CV is more appropriate for the PhD. The CV is used for academic positions and for some research positions in government and private laboratories.   If you are applying for a research position, it would be appropriate to ask whether a resume or a CV is preferred. Physicists are justly proud of their publications, but believe it or not, some employers are not interested in looking at a list of publications. Use your network to find out what is best for your particular case.

A CV tells the whole story; a resume tells the relevant story. The CV is many pages; the resume is, most typically, one page. For advanced degree holders, the resume can be two pages. The same CV can, more or less, be used for any job application; a resume is targeted to one job. When possible, however, a CV should also be tailored for a particular employer.

Since this book is about finding non-academic jobs, the focus will be on resumes.

After assembling the facts about your background, education, experiences, skills, and other professional data, you are ready to let the employer get to know you. The first step may seem easy, but it is not. You must sell yourself on one sheet of paper. The employer has no idea who you are, what you look like, and how you present yourself. The only thing the potential employer knows is 1) that your cover letter stimulated interest and 2) the information provided on your resume is, or is not, relevant to the company and the job position.

The purpose of the resume is to get an interview.

Each resume must be carefully crafted for the specific position being sought.

74

The resume is a brief presentation of your professional self: your education background, your skills/talents/knowledge *relevant to* the specific position you are hoping to land, and a glimpse of your aspirations. Preferably, all in one page.

Resume writing is an art. No matter how well prepared you are as a physics student, the skill to write an eye-catching and attention-getting resume must be mastered if you hope to get invited for an interview. Within the first few seconds, the reader of your resume will determine whether you deserve further consideration. Further consideration most likely means an interview. But wait, wait, wait. The initial "reader" may not be a real live person. It may well be a machine. (See Format of Resume below.)

Employers agree that resumes tailored to meet the specific requirements of a job are more attractive. Therefore, it is important to revise your resume each time you apply for a new job. Obviously, your education will remain constant for each revision; however, your statement of objectives, the skills you highlight from your educational and work experiences, and what you choose to draw attention to will change from resume to resume. Employers can receive hundreds of resumes for a single job opening. With this kind of volume, employers develop screening mechanisms to eliminate all but the most qualified candidates. The more clearly your resume matches the company's needs, the more likely you are to survive the preliminary screening process.

There are electronic databases where you can send your resume. Such a resume would be neither company specific nor job specific and is not advisable for the typical first-time job seeker. If, however, you have a skill and/or experience that would be picked up in a key word search, you could consider depositing your resume in a database.

Always use good quality paper.

**Keep a 1-inch margin
on all four sides.**

Always proofread carefully.

75

Format of the Resume - Scannable

The basic parts and the layout of the resume are described below. The purpose here is to point out that resumes are often not read by a warm-bodied person, but scanned into a database by a room-temperature machine. What are the implications of this mechanistic approach?

> **First**, your resume should contain no fancy decorations, no horizontal lines, no strange fonts - all of which may not scan accurately. Your resume should be clean copy with a standard font. Of course, your resume should be clearly organized and attractive looking, but no funny stuff.

> **Second**, after your resume is scanned, a computer search for key words may occur. If the key words are not found, your resume may get no further attention. Where do the key words come from? Probably from the same source that prompted you to apply for the job, namely, an announcement describing the open position, an advertisement, or someone in the company (your network!) who can tell about key criteria. **Look carefully at the job announcement and, to the extent possible, import the job-descriptive words in the announcement into your resume.**

Format of the Resume - Basic Types

There are three main types of resumes. The most common type, and most widely used, is the *chronological* resume which chronicles your educational and work history step by step. The *functional* resume emphasizes what you have done and what you can do rather than where and when you have gained your experiences, strengths, and skills. Sometimes a *targeted* resume is identified as a third type of resume; in fact, however, a functional resume is *de facto* a targeted resume as one should design a functional resume in terms of the requirements for a particular position. Finally, a *combination* resume brings together aspects of functional and chronological resumes.

> ## Every resume, regardless of the type, should connect with the job being sought.

For the first-time job seeker, what to do? The preferred type of resume depends on what you can bring to it. Earlier you were advised to do things that would attract attention to yourself. Well, here is where the rubber hits the road.

The Chronological Resume

A chronological resume documents your education, work experience, and relevant life experiences in **reverse** chronological order - the most recent dates first. It tells an employer exactly what you have done and when you did it.

If you have had summer jobs, if you have had an internship, if you have had teaching or a research assistantship, if you have done volunteer work or community service, a chronological resume is a good option. By contrast, if you have attended classes faithfully, but nothing else, a chronological list of relevant experiences might look rather skimpy.

The Functional Resume

A functional resume is organized around experiences and skills that can be targeted to a particular position - the position being sought! Functional resumes are better if you have had experiences and developed skills that you can describe and highlight. Here you can call attention to particular education- or work-related activities. If among other things, for example, you have

- had summer jobs,
- been a grader,
- done undergraduate research,
- written a dissertation,
- applied your "book learning" in some practical way,
- been a research assistant,
- been a teaching assistant,
- had an internship,
- developed computer skills,
- learned a second language,
- been a student leader in some ways,
- good writing skills,
- good speaking skills,

you can give prominence to these things (no chronological order) in a functional resume. Of course, link these experiences/skills to the job you are applying for; in fact, you might list them by order of importance as they relate to the desired job.

Combination Resume

A combination resume typically begins with a summary of your experiences and skills as you would do in a functional resume. This would be followed by a chronological record of employment and educational experiences.

Adopt a resume format that

displays your merits most prominently.

Format of the Resume - Basic Parts

The format of a resume is not set in concrete. The format of your resume should be chosen to highlight the strengths you offer your potential employer **in the context of the position you are applying for.** Your resume should connect with the job you seek. Here are the basic parts of the resume. Sample resumes follow.

Personal information

Name, address, phone and fax numbers, and e-mail address. This information should be at the top of the page.

Objective Statement or Summary Statement (Optional)

Objective: a one-sentence statement of your goals or what you are specifically looking for.
Summary: highlights your qualifications in one or two sentence(s).

Education

As a new graduate, the education section should be placed right after Personal Information or, if you elect to write a statement, right after either the Objective or Summary Statement. List your educational experiences in reverse chronological order. In this section include:

Name of institution and degree(s)
Address of institution
Year of graduation along with any academic honors (Cum Laude, etc.)
Major field of study, minor field of study

Do not include your grade point average (unless asked); do not include the name of your high school. If you are graduating with a PhD, do not include the name of your thesis advisor or the title of your thesis (unless your thesis research is directly related to the position you are applying for).

Experience/Work

   In this section, include three to five work-related or other experiences that serve to demonstrate your abilities, knowledge, and skills. These should be listed chronologically, with the most recent first. For each experience/work position cited include the name and location of employer, job title, and dates of experience/work.

For first-time job seekers, here are possible inclusions for your resumes:

- summer jobs
- internships
- teaching assistantships
- undergraduate research positions
- experiences that reveal computer-related skills
- experiences that demonstrate foreign language skills
- activities that demonstrate communication skills
- experiences in elective courses outside your major such as a writing course or business course
- campus or department leadership positions during undergraduate or graduate years

Other Possible Sections

   The following sections are optional, but each should be considered if it casts you in a good light. For example,

- Awards, honors, scholarships, fellowships
- Volunteer activities
- Memberships in professional-related organizations

DO NOT INCLUDE

   1. By law, employers are not permitted to ask personal questions such as your age, your marital status, or the number of children you have, so do not include this information in your resume.

   2. Do not feel compelled to fill empty space at the bottom of your resume with hobbies or other personal activities. Such information is not needed by your future employer.

   3. **References** - In general, references should not be listed on your resume. If the employer wants references with the job application, it will be so indicated (for example, in the

job advertisement). If references are requested, identify your references on a separate sheet of paper with name, job title, place of employment, relationship to you, address, phone number, fax number, and e-mail address.

Whom do you ask for references?

- Someone who knows something about you beyond the classroom and can say more than, "Sally took my course in Electricity and Magnetism."

- Someone who you have had a good personal relationship with. **Note:** This may exclude your thesis advisor. Think carefully about this.

What do I do for the person writing a reference for me?

- Tell the individual you would like to use her/him as a reference.

- Provide the reference writer with a brief description of the position you are applying for.

What catches my eye about a resume is:

- **combination of theoretical and laboratory experience,**
- **independent experience and experience in working as a member of a team,**
- **interests outside physics - do they have a broad set of interests,**
- **system engineering understanding, and**
- **experience with data acquisition/analysis systems**

David Robson, Xerox

Resume Writing Tips

Although the tips listed below may appear repetitious, these ideas are specifically geared to technical resumes.

1. Use as many "buzzwords" as you can to reflect your work and school experience. For example list all operating systems and special applications such as UNIX and SPSS with which you have experience. Use key words that appear in the job advertisement.

2. Begin sentences with action verbs (see list below). Portray yourself as someone who is active.

3. Don't sell yourself short. Treat your resume as an advertisement for your abilities.

4. Be specific about your accomplishments and give them emphasis.

5. Don't exaggerate your qualifications.

6. Be concise. Avoid lengthy descriptions.

7. Always use correct grammar and spelling. Have a friend proofread.

8. Use high quality bond paper.

**Physicists are particularly well suited
for interdisciplinary research
because of their thorough grounding
in fundamental concepts
and the power and broad applicability
of their theoretical, experimental,
and computational methodologies.**

Kenneth C. Hass, Ford Research Laboratories

I went to a large career fair that was part of the national Society of Women Engineers (SWE) convention in Chicago. I sent my first round of resumes out to those companies. I also posted my resume on the SWE Resume Database. However, the avenue that produced my first job offer from the Missouri Department of Transportation (MoDOT), my first employer, was to go through my university's placement office. I signed up for an interview when MoDOT came to campus. I got a job offer. I would consider collegiate career placement centers and university sponsored career fairs to be the best opportunities for college seniors.

Sharon A. Lappin, P. E.
Project Engineer
Missouri Department of Transportation

Curriculum Vitae

As stated earlier, a curriculum vitae (CV) is mainly used when applying for research positions. CVs are always used in the academic world. CVs are also used when applying for educational jobs, research jobs in government or private laboratories, or for a fellowship or grant. Like a resume, a CV is a summary of your educational and professional background that makes you qualified for a particular job.

Unlike your resume, which is typically one page, CVs are several pages in length and should include a full list of your publications, honors, awards, committee work, memberships, and other professional connections and experiences.

The format of a CV is the same as a resume discussed above. The same sectional organization can be used. In addition, however, there is a final section where publications are listed in reverse chronological order. If you have several publications, you could organize them in subsections: peer-reviewed publications, in-press publications, and conference proceedings.

If your research has attracted attention and you have given invited talks, you should have a section, just before the section on publications, where you list invited talks.

As with the resume, references go on a separate page.

> **Many PhD resumes just scream "I have no idea what your business environment might be; I expect to do more of my thesis work and it's up to you to extract the value." It is critically important that PhDs demonstrate excellent communication skills, and an understanding of the business context of the company to which they are applying, or they can be pigeonholed. Please note that these considerations apply even at places like Bell Labs, IBM Watson, etc. not just "dirt under your fingernails places."**
>
> John Sommerer, Director, Research and Development Technology
> Johns Hopkins Applied Physics Laboratories

Action Verbs for Your Resume

There are dull words and exciting words. The former can put a reader to sleep, the latter can grab the reader's attention. Action verbs are action words and, when used in a resume, they represent you as an active person. Action verbs are recommended. Below are action verbs for all occasions.

ACTION Verbs . . .

. . . to express Accomplishments

Achieved	Expedited	Pioneered
Attained	Improved	Resolved
Completed	Increased	Restored
Convinced	Initiated	Revitalized
Doubled	Introduced	Revolutionized
Discovered	Invented	Spearheaded
Eliminated (costs)	Launched	Strengthened
Expanded	Originated	Transformed

. . . to express Communication skills

Addressed	Edited	Persuaded
Arbitrated	Enlisted	Promoted
Arranged	Formulated	Proposed
Adhered	Influenced	Publicized
Collaborated	Interpreted	Reconciled
Corresponded	Lectured	Recruited
Developed	Mediated	Spoke
Directed	Moderated	Translated
Drafted	Negotiated	Wrote

. . . to express Creative skills

Acted	Fashioned	Revised
Conceptualized	Illustrated	Revitalized
Created	Instituted	Set up
Customized	Integrated	Shaped
Designed	Performed	Simplified
Directed	Planned	Streamlined
Established	Proved	Structured

> **Do not verbalize nouns. Strategy is a noun; strategize is a verbalized noun and is not a word. Many people take offense at such a practice.**

. . . to express Detail skills

Approved	Generated	Purchased
Arranged	Implemented	Recorded
Catalogued	Inspected	Retrieved
Classified	Monitored	Screened
Collected	Operated	Simplified
Compiled	Ordered	Specified
Dispatched	Organized	Tabulated
Executed	Prepared	Validated
Filed	Processed	Verified

. . . to express ability to Help others

Assessed	Familiarized	Rehabilitated
Assisted	Guided	Represented
Clarified	Inspired	Reinforced
Coached	Led	Supported
Counseled	Motivated	Taught
Demonstrated	Participated	Trained
Diagnosed	Provided	Verified
Educated	Referred	

. . . to express Management skills

Administered	Developed	Produced
Analyzed	Directed	Recommended
Assigned	Evaluated	Reorganized
Chaired	Led	Revamped
Consolidated	Organized	Reviewed
Contracted	Oversaw	Scheduled
Coordinated	Planned	Supervised
Delegated	Prioritized	

. . . to express Research skills

Clarified	Evaluated	Interviewed
Collected	Examined	Investigated
Critiqued	Extracted	Organized
Diagnosed	Identified	Reviewed
Discovered	Inspected	Summarized
Elucidated	Interpreted	Surveyed

. . . to express Teaching Skills

Adapted	Enabled	Lectured
Advised	Encouraged	Persuaded
Clarified	Evaluated	Presented
Coached	Explained	Set goals
Communicated	Facilitated	Stimulated
Conducted	Guided	Taught
Coordinated	Informed	Trained

. . . to express Technical skills

Assembled	Devised	Pinpointed
Built	Engineered	Programmed
Calculated	Fabricated	Remodeled
Computed	Maintained	Repaired
Designed	Operated	Solved

Whate'er you think,

good words, I think,

were best.

William Shakespeare

Proper words in proper places,

make the true definition

of a style.

Jonathan Swift

Sample Resumes/CVs

Chronological Resume

Robert Goodlong   TEL: (324) 648 4698
137 Fine Structure St.   FAX: (324) 648 5285
Arlington, VA   Email: rgoodlong@abc.org

Career Objective (See Note 1)
Outline your career objective (relate it to the available position).

Education (See Note 2)
9/00-6/02   MS Physics, Sommerfeld University

9/96-6/00   BS Physics, Sommerfeld University (*magna cum laude*)

Experience: (start with most current experience/job)

9/01-6/02   Research Assistant, Department of Electrical Engineering, Sommerfeld University
Briefly describe the scope and responsibilities of the job you
currently hold. Be sure to include any major accomplishments that
demonstrate your suitability for the job for which you are applying.
You may use bullet or paragraph format.

9/00-6/01   Teaching Assistant, Department of Physics, Sommerfeld University
Again, briefly describe the scope and responsibilities of the
position. Emphasize those accomplishments that relate to the job
for which you are applying.

6/00-9/00   Internship, Digital Atomics, Falls Church, VA
Include a sentence describing your responsibilities at Digital Atomics.

Other Accomplishments
List honors or awards outside of educational or other
achievements. You may also want to list memberships in
professional organizations. Be brief.

Note 1: If your career objective connects rather directly with the position you are applying for, a career objective statement is an option to consider.
Note 2: Either education or experience can be the lead entry. If you have experiences that have prepared you for the position you seek, lead with experiences rather than education.

In a resume do not:

~ **include name of your high school**

~ **include salary information**

~ **include personal information (marriage status, family details, religious interests)**

~ **include names of references**

Chronological Resume

Name
Address
City, State, Zip
Phone Number
E-mail address

Objective

Insert the type of work you want to do. Customize for each job opening.

Experience

COMPANY/LABORATORY NAME

Job Title, Department/Division, Dates

  *Emphasize results, accomplishments and performance.

  * Don't list only job description and duties - show how you made a difference.

  * List each accomplishment statement in order of importance.

  *Use an action verb as first word for each "bullet".

  *Keep to one or two lines.

  *Be honest. Be positive. Be specific. Be brief.

  *Be quantitative whenever possible, i.e., increased by 30%.

Job Title, Department/Division, Dates

  *Omit detailed descriptions of nonrelevant earlier jobs.

  *Use a consistent format.

  *Account for lengthy time gaps (i.e., school, pregnancy/baby, self-employed).

  *Two pages maximum (advanced degree). Attach publications separately.

COMPANY/LABORATORY NAME

Job Title, Department/Division Dates

  *Keep a one-inch margin on all four sides.

  *Print on top-quality resume paper - white, cream or gray.

  *Proofread until there are absolutely no errors; have several other people proofread it as well.

Education and Training

List in the following order: degree, major, school, city and state, year of graduation. If it is recent and relevant to the position, put this section before "Experience"; if not, include here. Include any training that supports the job you seek. Incorporate any other credentials or certificates.

Optional: Technical skills, military service, awards, professional associations.

Functional Resume

Name
Address
City, State, Zip
Phone Number and e-mail address

Professional Summary: Briefly describe your employment history, highlighting key accomplishments or areas that you want to sell to a prospective employer.

Areas of Effectiveness (Be very specific in this listing)
Number 1 skill:

Latest and Best Example of use of this skill. Can include one assignment or part of your current or last job.
Next Best Example of use of this skill. Can be made up of a job element, an assignment, or major accomplishment.
Next Best Example of use of this skill. Can be an older example, a significant accomplishment, or achievement.

Number 2 skill:

Best **Example** of use of this skill. Again, can be made up of single events, assignments or parts of jobs that you enjoyed or did well.
Next Best Example of use of this skill. Earlier, older or less significant events.

Number 3 skill:
Best Example of the use of this skill. Sometimes involves hobbies, nonpaid work activities, or may be derived from less important work skills.

Employment History
Latest or current job:	Title, Organization & Dates
Previous Job:	Title, Organization & Dates
Previous Job:	Title, Organization & Dates

Education
Begin with highest and latest degree OR educational accomplishment, then work backward. List DEGREE / DIPLOMA, SCHOOL, DATE.

Other Professional Accomplishments
Honors or awards, superior achievement, memberships in professional organizations, and the like.

Functional/Skill Resume

Name
Address
City, State, Zip
Phone Number
e-mail Address

Objective

This statement answers the question, "What are you seeking?" In one or two sentences, address the type of position, title, or area you seek. Include the level of responsibility and two or three personal characteristics or skills needed on the job you have targeted.

Summary

Include a descriptive summary of yourself in three or four sentences. This should contain how many years of experience you have, the environment in which you worked, your areas of expertise, and brief descriptions of your most salient professional or scientific characteristics. This should give a quick image of your overall qualifications and invite the employer to continue.

Areas of Accomplishment

Key Skill

* Cluster your experience under major skills areas.
* Incorporate your strongest skill first.
* Focus on accomplishments in functional or technical areas.
* Select and organize key skills to support your objective statement.

Key Skill

* Describe your acquired capabilities that match with the employer's job qualifications.
* Feature the skills that are essential to succeed on the desired job.
* Emphasize transferable skills that relate to the position you seek.
* Refer to the action verbs and adjective list.

Key Skill

* This resume should be used when changing careers, or you are trying to explain an erratic employment history.
* Functional resumes may be used when you wish to increase your level of responsibility.
* Use this format when you want to expand your breadth and have an unusual combination of functional areas and skills.

Employment History

2001 - present Job Title   Place of Employment

Education and Training

List in the following order: degree, major, school, city and state, and year of graduation. If it is recent and relevant to the position, put this section before experience; if not, include here. Include any training that supports the job you seek. Incorporate any other credentials or certificates.

Optional: Technical skills, military history, awards, professional associations.

Chronological CV

EDWARD T. PHYSICIST

Institute for World Science
University of Today
City, State 12345
210-555-5555 (work)
210-999-9999 (home)
ephysicist@iws.org

OBJECTIVE

To perform research and analysis working as an environmental scientist.

EDUCATION

PhD Physics, State University of the Sciences, City, State, Thesis Completion (Anticipated) June 2002, (Thesis Title ?)
MS Physics, State University, City, (Thesis Title ?), December 1999.
BS Physics, WW University, City, Phi Beta Kappa, June 1997.

EXPERIENCE

8/99 - present: Research Assistant, State University. Computational work involving the modeling of the night side ionospheres of Venus and Mars, including data analysis of Pioneer Venus orbital data.

9/99 - 6/01: Lecturer, Department of Physics and Astronomy, State University. Instructed senior-level astrophysics courses.

8/97 - 8/99: MS Physics Research, State University. Measured and analyzed infrared spectral lines of CH_2, $C_2 H_4$, and NH_3 broadened by various perturbing gases, to determine spectral linewidths that are of interest in the modeling of planetary atmospheres.

9/95 -6/97: Lecturer and Lab Instructor, Department of Physics and Astronomy, WW University, Wacamaw. Taught the freshman honors astronomy lab; served two summers as teaching assistant for the Wacamaw Governor's School for the Sciences physics group.

COMPUTER SKILLS

Proficient with DEC VAX (VMS operating system), and IBM PC, with knowledge of FORTRAN, Quick Basic, and LATEX.

OTHER EXPERIENCE

Astronomy Programs. Conducted and assisted with astronomy observations and lectures for area school systems, clubs and the public during graduate work at State University. Served as a judge for the State University regional Science Olympiad astronomy contest for two years.

6/94 - 9/94: Photo Darkroom Technician, News-Herald, Broadway, City/State. Developed and processed photographs for county newspaper. 6/82 - 6/83.

Customer Service, Meeger Co. and Hi-Lo, Little One, Wacamaw. Worked for two grocery stores during undergraduate studies.

RECENT PUBLICATIONS

"Arguments for Day-to-Night Transport at Low Solar Activity," T. J. Remington, Edward T. Physicist, P. H. Longbrow, and H. L. Streamer, *Physical Reporter* **13**, 523-529 (2000).

"Upper Limits to the Night Side of Mars," T. J. Remington, Edward T. Physicist, P. H. Longbrow, and H. L. Streamer, *Physical Reporter* **12**, 145-151 (1999).

RECENT PRESENTATIONS

"The Nightward Fluxes of 0 in the Venus Ionosphere," T. J. Remington, Edward T. Physicist and P. H. Longbrow, World Association of the International Organization of Physics and Geophysics, Buenos Aires, Argentina, August 1993.

"Model Calculations of the Nightside Ionosphere of Venus and Mars," T. J. Remington, Edward T. Physicist, P. H. Longbrow, J. J. Evans, and Y. O. Ewe, Annual Meeting of the Division for Planetary Sciences of the AAKI, Munich, Germany, October 1990.

Invited Paper

"The Nightside Ionosphere of Venus," T. J. Remington, Edward T. Physicist, J. J. Evans, and Y. O. Ewe, Meeting of Pioneer Planetary Commission, Wright Research Center, Kato Court, California, presented by J. J. Evans, September 1991.

SCHOLARSHIPS AND AWARDS

Coblentz Society Student Award, 1987
Wal-Mart Foundation Scholarship, 1986
University of Wacamaw Alumni Valedictorian Scholarship, 1986

PROFESSIONAL MEMBERSHIPS

Member, American Physical Society
Member, American Astronomical Society
Member, American Association for the Advancement of Science

THE INTERVIEW

> **Step 9**
> **Prepare for your interview.**
> **Know the employer, know the position,**
> **know how to connect**
> **your talents with the position.**

The interview is the critical step along the path to employment. It determines whether you will be offered a job or dismissed from the applicant pool. In addition to careful preparation for the interview, there are common-sense guidelines (rules?) to follow before, during and after the interview.

Let's start at the beginning.

> **The cover letter and resume**
> **get you the interview.**
> **The interview gets you the job.**

> **Getting an interview**
> **means you are at**
> **or near the top of their list.**

> **The purpose of an interview**
> **is to get a job offer.**

Before the Interview

Prepare. There is no way to overemphasize the importance of careful and thorough preparation. Know the company, understand the company culture, know the job requirements, know how the job is situated in the context of the company. Prepare, prepare, prepare.

Be prepared to talk about yourself and your goals. Make a list of important background information that you want to convey in the interview. Be sure to include your academic and non-academic accomplishments. Also, be prepared to discuss both your long-term and short-term professional goals with the interviewer.

If possible, know the type of interview in advance. There are different types of interviews for different purposes. Prepare as best you can for the type of interview you will encounter. You may well have more than one interview during a day or a week. Below are different types of interviews you may encounter.

Screening Interview. These are used when colleges and universities hold on-campus interviews, or a company has a substantial number of people applying for a job. A screening interview is designed to help employers sort out the applicants. These interviews are short, 20-30 minutes, and basic questions are asked such as: "What makes you qualified for this job?"

One-on-One Interview. This is the most common type of interview and it is exactly as it sounds: just you and one other person.

Phone Interview. A phone interview is sometimes used when time and cost are issues. As with any interview it will be scheduled in advance. Arrange to have your phone interview in a quiet place away from barking dogs, noisy children, or other distractions. Maintain a degree of formality in your questions and answers,

Panel/Committee Interviews. In a panel interview there will likely be several interviewers present. Instead of directing your answers to one person, you have to engage everyone (eye contact!) as you are responding to questions. Panel members may interrupt each other and you. Keep your poise and follow through on your answers. This type of interview is not common for entry-level positions.

Case-Study Interviews. Again, not common for entry-level positions. In a case-study interview, a concrete situation will be presented and you are asked to lay out how you would respond. The situation presented will likely be one you are unfamiliar with. This approach is used to evaluate your approach to problem-solving, evaluate your analytical abilities, and to see how you arrive at a logical conclusion.

Stress Interviews. If you are going through a series of interviews, this type may be part of the series. These are used to see how you handle pressure. Difficult questions may be asked, you may have an impatient interviewer, or an interviewer who deliberately tries to destabilize you. This type of interview is unlikely for an entry-level position.

**Slick, slangy talk does not
do the job
they are interviewing you for.**

Common Interview Questions. Interview questions can certainly vary; however, there are some common questions. Here are a sampling of questions you may be asked during an interview. Some of these, or questions like them, *will* be asked. Think about these questions, consider how you would respond, and practice your response.

1. What interests you about this job?

2. How did you become interested in this field?

3. What two or three things are most important to you in a job?

4. What aspects of our organization/company interest you?

5. What do you think it takes to be successful in an organization like ours?

6. What do you see yourself doing five years from now?

7. How do you determine or evaluate success?

8. Are you willing to relocate? Any geographical preferences?

9. Why should I hire you?

10. What have you done that is applicable to this position?

11. This position requires that you _____ . Describe in detail how your background experience will enable you to do this?

12. In what ways do you think you can make a contribution to our company?

13. What is your greatest strength? Greatest weakness?

14. How would you describe yourself?

15. How do you think a friend or professor who knows you would describe you?

16. What is most important to consider when dealing with _____?

17. If _____ occurs within this position, what would you do?

18. In what kind of environment would you be most comfortable?

19. Tell me about a problem you experienced and how you handled it. In retrospect, how would you improve on that?

20. What things do you find difficult to endure?

21. Tell me about a conflict you had with another person and how you dealt with it. Is this characteristic of how you generally approach conflicts with people?

22. How do you manage stress?

23. Can you work under pressure? Are you able to manage your time effectively?

24. How do you handle criticism?

25. Are you willing to admit to your mistakes?

26. Tell me about a mistake you made in the past and what you learned from it.

27. If you disagree with something your boss told you to do, what would you do?

28. Describe the relationship that should exist between a supervisor and subordinates.

29. What do you expect to be earning in five years?

30. Which is more important to you, the money or the job?

31. Do you have any questions?

Dress appropriately. If you have prepared properly, you know how employees typically dress. Do not dress as employees do, that is, as if you are going to work. You are not going to work. Do not wear a sweatshirt and sneakers. Do not dress like your professors.

At the same time, do not dress 25 times fancier than the standards of the company employees. As a general rule, a jacket/tie for men and a skirt/dress or pants/jacket for women is recommended. Some would say that you should always wear a suit to any interview, and you can follow that line. Here is where knowing about the corporate culture is important. You want to convey the impression that you will fit in. In a casual dress environment your interviewer may be wearing jeans. If you appear in a three-piece suit, both you and the interviewer may feel awkward and uncomfortable. Not too dressy, not too casual. You do not want your appearance to call attention to itself.

Be punctual. In fact, arrive a few minutes before the scheduled time of the interview. Know where the interview is scheduled, building and room number; know how to get there and where to park. To say, "I am sorry, I got lost", is not a good beginning for an interview. Plan to arrive early, get a cup of coffee in the lunchroom, find the restroom, straighten your hair, and arrive at the designated location breathing normally.

Bring appropriate materials. Bring your resume and any other materials that reflect your work talents and experience. Be prepared to provide the interviewer with referral names on the spot.

Be prepared to ask questions. Even if you are well informed about the company and certain that you want the job, always ask the interviewer a few related questions. This demonstrates your interest in the company and gives the interviewer a chance to do a little talking.

Compensation. Suppose the interview goes so well that you are offered a job on the spot. With the job offer, a salary figure may well be identified. What do you do? See Negotiating a Salary in the section to follow.

> **The best advice I can give anyone at an interview is - look your best and relax. Just answer the questions honestly and completely.**
>
> Larry Collette

During the Interview

After introductions, ask for the business cards(s) of all those present. It is appropriate to ask each person present what his/her position is in the company. If possible, make connections in your answers with particular positions.

Important hint. The first few moments of the interview can be decisive. First impressions are important; first impressions are remembered.

Listen to the question before answering. Answer the interviewer's questions directly. It is okay to discuss other issues and experiences, but lead into them after you answer the question.

Demonstrate interest. If you recognize that you are qualified for and interested in the job, demonstrate this by projecting images to the interviewer that emphasize your knowledge and expertise in the field. Talk specifically about your relevant past experiences. If you know how to use an unusual piece of equipment that connects with the position, describe your experience and show how your expertise could be applied in the desired job. Project your knowledge, understanding, and abilities. Those interviewing you will be looking for a match between you and the organization. As they are asking questions, they will be wondering what skills, knowledge, and experience you can contribute. Also, do not forget this is your interview. Ask questions about the work environment because you have already assessed your needs and you know what working conditions you are looking for.

Pay attention to your body language. Your expressions and physical gestures speak to the interviewer about your attitudes and personality. There are some basic symbols of confidence that you want to be sure to convey. Below are some tips about body language.

Start off with a firm handshake. Look the interviewer directly in the eye, stand up straight, and walk confidently into his or her office. Do not fidget during the interview.

Facial Expression. Be serious, yet friendly. This means you should smile during introductions and small talk, etc. However, maintain a more serious expression as you discuss your skills and respond professionally.

Eye Contact. Make eye contact with the interviewer. When the interviewer is asking a question, make eye contact. When you are answering a question, make eye contact. Do not look down at your watch or at the clock. Do not appear distracted by other people in the office.

Hand and Feet Movements. Sit still. Tapping your fingers or fidgeting shows that you are nervous. Do not sit with your legs crossed or with your feet crossed. Try to refrain from crossing any parts of your body as this could be interpreted that you are not an open person and are trying to hide something.

Voice Variations. Try to keep your voice from wavering. Use appropriate inflections in pitch. Avoid the monotone. Speak slowly and confidently.

Silences. Avoid long silences, but don't be afraid to pause briefly between question and answer. This shows deliberation and gives you time to construct your response.

Posture. Stand and sit straight. This projects attentiveness and self-confidence.

Presenting your research during an interview. During a second or third interview, an employer will often ask you to give a presentation of your significant research project(s) to a panel. This is may be a screening device in the later stage of the hiring process. Below are some tips for effective presentations.

Plan Your Presentation. If possible, determine who will be in the audience and how much time is provided for your talk. Stay within the time limits. Ask yourself, "What in my research will be important to the members of the audience and to the organization?" Capture the essence of your research in a small number of main topical areas. Develop your talk from each of these main topics, concentrating on your accomplishments. Tie your research accomplishments to the needs of the organization, i.e., its strategic mission, or the products it is interested in developing. Have handouts available if appropriate for clarity. Be sure to summarize your main points at the conclusion of the presentation. This will help to reinforce these points for better retention by the audience.

Delivery. Practice your presentation carefully in front of a friend, or record yourself on video and review the tape. Visualize success. Be animated, enthusiastic and natural. Be aware of what you are saying and speak conversationally with a strong clear voice. If asked questions, maintain your style, repeat the question, and involve the entire audience in the answer. Remember to reserve 25 percent of your eye contact for the person who asked the question. Focus on relaxing - release tension by contracting and releasing hand muscles, try deep breathing. Stand, if possible, because it will give you more audience control. Maintain eye contact with your audience.

Questions that should not be asked. During an interview there are questions that the interviewer should not ask. You have the legal right to refuse to answer these questions as they may be grounds for discrimination against you.

1. Your religion, political beliefs, or affiliations.

2. Your ancestry, national origin, or parentage.

3. The naturalization status of your parents, spouse, or children. It is permissible to be asked whether or not you are a U.S. citizen, or the status of your visa.

4. Your birthplace.

5. Your native language (but you can be asked about the languages your resume claims you speak).

6. Your age, date of birth, or ages of children (but you can be asked whether or not you are the age of 18).

7. Your maiden name, or whether you changed your name, your marital status, number of children or spouse's occupation, or anything about your spouse. (The last is the most commonly encountered illegal question asked of female job applicants.)

During the interview, if any of these questions are asked (or others that you feel are inappropriate), do not act offended. You may just want to ask, "Is this question relevant to the position for which I am applying." You may also politely refuse to answer the question.

Be yourself, be confident. Show the interviewer that you are motivated and want to do well. No one wants to hire a mediocre employee. Most important, be yourself. Do not attempt to be what you imagine the interviewer wants you to be. It could come off as a lack of genuineness. Relax and speak in your natural manner. Use your common sense and let the interviewer witness your personality. Look at the interview as a presentation - a presentation of yourself that should be substantive and interesting. Do not just recite data, tell a story. Be enthusiastic. Be clear. Be brief.

After the Interview

Express your appreciation to the interviewer(s). The simple act of saying "Thank you" cannot be over emphasized.

Send letter(s) of appreciation. You have the business cards of everyone you spoke with. Within a day or two after the interview, send a thank you note to each of them. Briefly restate your qualifications and your interest in the position. Make your letter short.

PHYSICS TRENDS

Typical Starting Salaries for Physics Bachelors, Classes of 1999 & 2000

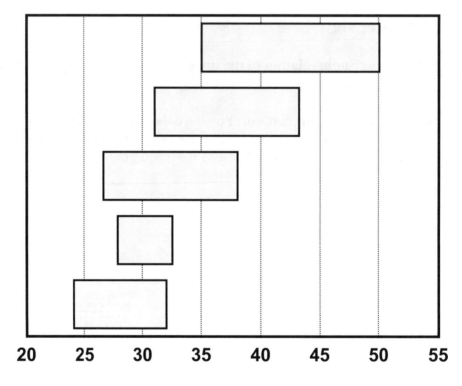

Employer

Industry

Civilian Government

College or University

High School

Active Military

Salary in Thousands of Dollars

Source: Initial Employment Survey, Classes of 1999 and 2000

AMERICAN INSTITUTE OF PHYSICS

Statistical Research Center
www.aip.org/statistics

PHYSICS TRENDS

Typical Salaries for PhD Physicists Early in Their Careers, 2000

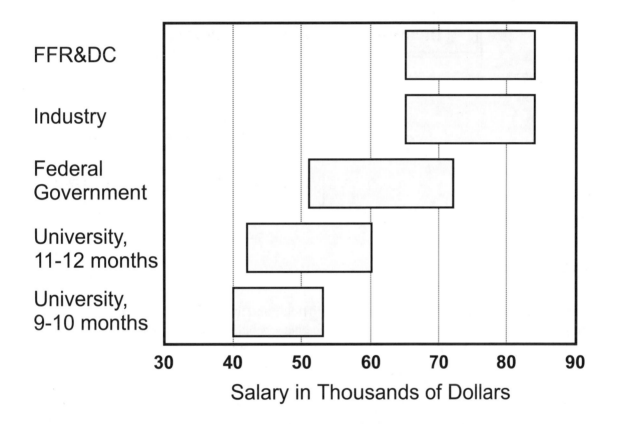

FFR&DC

Industry

Federal Government

University, 11-12 months

University, 9-10 months

30   40   50   60   70   80   90

Salary in Thousands of Dollars

Typical salary ranges for PhDs who have worked for less than five years

FFR&DCs are Federally Funded Research and Development Centers.

Some university salaries are for a full year of work, while others are paid only for the 9 to 10 month academic year.

Source: 2000 Salary Report

AMERICAN INSTITUTE OF PHYSICS

Statistical Research Center
www.aip.org/statistics

NEGOTIATING COMPENSATION

Compensation = Salary + Benefits

The longer a company courts you, the more valuable you become.

Do not discuss salary until you receive a firm job offer.

Starting Salary

For most people, talking about salary is difficult and awkward. This is why agents make a good livelihood. It is easy to argue on someone else's behalf; it is hard to argue for yourself. However, being prepared to discuss salary, and doing so knowledgeably, demonstrates for the employer that he/she made a good decision to hire you.

Before you go for an interview, it would be wise to give compensation careful thought. To begin this process, you should find out the salary range for a position such as you are seeking. Go beyond the particular employer you will be interviewing with. Do a search on the Internet where you can find salary surveys. Classify your potential employer in terms of type: government, private sector, large corporation, small high tech firm, etc. Classify by region. In terms of your classification, determine what similar employers typically pay - high, average, low - for positions with equivalent job requirements.

Next, in light of the job requirements, assess your qualifications as accurately as possible and decide where in the range you believe you fall. Decide in advance the lowest salary you will accept. If the employer asks you to suggest a salary, you could answer with a question of your own. For example, "What is the salary range for the position you have offered me?" In short, try to avoid being the first to identify a specific salary. If you must name a salary figure, give a figure larger than your minimum salary and be prepared to provide a rationale for your suggested figure.

> **If an employer can "get you" for $1,000/year, he/she will not offer you $1,001/year.**

If the employer counters with an offer that is below the level of your request, remember, it is the obligation of an employer to hire you at the least possible cost. The first offer to you will likely be a lower figure than he/she is willing to pay. If after discussion, the offer still is lower than you want to accept, you might compromise by asking for a three month salary review on the condition that if your work is up to standards, your salary will increase to your requested level.

> **The starting salary is the base you start from. The larger the base, the larger the salary increases will be forever and ever.**

Salary Increases

As a part of salary discussion, ask about how salary increases are determined and how frequently they are given. Annual salary increases are typical. How are salary increases divided between cost of living increases and merit increases? Are merit salary increases based on your job description?

While an employer's policy concerning salary increases is important, your starting salary is the base you work from. Do a sample calculation. Assume a 4% annual merit salary increase, for 10 consecutive years, starting from two different salaries, say, $45,000 vs. $48,000.

The Five Keys to Salary Negotiations

1. Never discuss salary till the end of the interviewing process, when they have definitely said they want you.

2. Never be the first to mention a salary figure.

3. Before you go to the interview, do homework on how much you need.

4. During the interview, try to determine whether the salary being offered is fixed or contains room for negotiation.

5. Before you go to the interview, do research on salaries for your field or for that organization.

Richard Bolles, *What Color Is Your Parachute?*

Benefits

Standard Benefits

Do not underestimate the importance of your benefits. Benefits can amount to 25-50% of your base salary. It is money into or out of your pocket. Here are benefits that are offered by most employers:

Time off - Time off = vacation + national holidays + personal days.

Vacation time will be offered, but is it ten days or three weeks?

What national holidays are given as time off? (9 to 12 days are typical.)

Are personal days given (three days/year is common)?

If you observe non-Christian religious holidays, what is company policy?

Health Insurance - Health costs are substantial, so health coverage is important.

What kind of coverage is provided?

How are the costs shared between employer and employee?

Is dental coverage included?

Is vision coverage included?

Disability and Worker's Compensation - If you are injured on or off the job, some fraction of your salary continues. The employer pays the premium.

Maybe Yes, Maybe No

Retirement Plans - Most retirement plans are shared between employer and employee. The question is: How much does the employer contribute? A 10% contribution is ballpark.

Tax deferred plans - Plans such as 401(k) siphon part of your salary off the top, no taxes until retirement, and put it into a savings program. Employers sometimes make a contribution along with the employee's.

Day Care Services - Some companies have day-care centers.

Flextime - Starting and quitting times can be adjusted to one's convenience. Total time on the job is still full-time.

Profit Sharing - Usually some percentage of base salary.

Car Expenses - If your job description calls for driving your car, the employer should have a program to cover your associated expenses.

Dues for Professional Societies - If you belong to a professional organization that relates to your job, some companies will pay your dues.

Tuition Expenses - If you need additional course work, many employers will pay the tuition costs.

After an Offer Is Accepted

Once you have accepted a job offer, there are a few things you should do.

Establish the parameters of the workday - starting and quitting times. If you are to be a salaried employee, starting and quitting times are not relevant. You get the job done - 8 hours/day or 12 hours/day.

You have been hired to for a particular job. Ask for a job description.

Talk to your new boss. Find out what he/she expects for your first six months. Together, set some goals.

Ask if there is an employee handbook. It will describe many useful things.

Ask for a company organization chart. It will reveal how your job fits into the larger corporate picture.

> **Two roads diverged in a wood, and I -**
> **I took the one less traveled by,**
> **And that has made all the difference.**
>
> Robert Frost, from "The Road Not Taken"

Your decision to pursue a degree in physics put you on a road "less traveled by." The road stretches long before you. Your first job is the beginning of what will become your career. Enjoy your new job.

1-MONTH